CELL BIOLOGY RESEARCH PROGRESS

MITOGEN-ACTIVATED PROTEIN KINASES (MAPKS)

ACTIVATION, FUNCTIONS AND REGULATION

CELL BIOLOGY RESEARCH PROGRESS

Additional books and e-books in this series can be found on Nova's website under the Series tab.

CELL BIOLOGY RESEARCH PROGRESS

MITOGEN-ACTIVATED PROTEIN KINASES (MAPKS)

ACTIVATION, FUNCTIONS AND REGULATION

CHARLES K. HESTER
EDITOR

Copyright © 2019 by Nova Science Publishers, Inc.

All rights reserved. No part of this book may be reproduced, stored in a retrieval system or transmitted in any form or by any means: electronic, electrostatic, magnetic, tape, mechanical photocopying, recording or otherwise without the written permission of the Publisher.

We have partnered with Copyright Clearance Center to make it easy for you to obtain permissions to reuse content from this publication. Simply navigate to this publication's page on Nova's website and locate the "Get Permission" button below the title description. This button is linked directly to the title's permission page on copyright.com. Alternatively, you can visit copyright.com and search by title, ISBN, or ISSN.

For further questions about using the service on copyright.com, please contact:
Copyright Clearance Center
Phone: +1-(978) 750-8400   Fax: +1-(978) 750-4470   E-mail: info@copyright.com

NOTICE TO THE READER

The Publisher has taken reasonable care in the preparation of this book, but makes no expressed or implied warranty of any kind and assumes no responsibility for any errors or omissions. No liability is assumed for incidental or consequential damages in connection with or arising out of information contained in this book. The Publisher shall not be liable for any special, consequential, or exemplary damages resulting, in whole or in part, from the readers' use of, or reliance upon, this material. Any parts of this book based on government reports are so indicated and copyright is claimed for those parts to the extent applicable to compilations of such works.

Independent verification should be sought for any data, advice or recommendations contained in this book. In addition, no responsibility is assumed by the Publisher for any injury and/or damage to persons or property arising from any methods, products, instructions, ideas or otherwise contained in this publication.

This publication is designed to provide accurate and authoritative information with regard to the subject matter covered herein. It is sold with the clear understanding that the Publisher is not engaged in rendering legal or any other professional services. If legal or any other expert assistance is required, the services of a competent person should be sought. FROM A DECLARATION OF PARTICIPANTS JOINTLY ADOPTED BY A COMMITTEE OF THE AMERICAN BAR ASSOCIATION AND A COMMITTEE OF PUBLISHERS.

Additional color graphics may be available in the e-book version of this book.

Library of Congress Cataloging-in-Publication Data

ISBN: 978-1-53616-138-0

Published by Nova Science Publishers, Inc. † New York

CONTENTS

Preface		vii
Chapter 1	P38 and JNK as Targets of Protozoan Parasites to Manipulate Host Immune Response and Survive Inside Host Cells *Laila Gutiérrez-Kobeh,* *Jorge Rodríguez-González,* *Rosalino Vázquez-López* *and Arturo A. Wilkins-Rodríguez*	1
Chapter 2	Roles of MAPKs in Circadian CLOCK Regulation in Vertebrates *Yoshimi Okamoto-Uchida, Junko Izawa* *and Jun Hirayama*	29
Chapter 3	Toxicological Significance of MAPK Activation in Cadmium-Induced Kidney Cell Death *Masato Matsuoka*	55
Bibliography		75
Index		91
Related Nova Publications		97

PREFACE

Mitogen-Activated Protein Kinases (MAPKs): Activation, Functions and Regulation opens with a summary of the present knowledge about MAPK, with special emphasis on p38 and c-Jun N–terminal kinase. The authors focus on how these signaling pathways are engaged during some infections with intracellular parasites. The authors also describe selected regulatory aspects of circadian clocks in vertebrates, exploring an intriguing link to MAPK. Circadian clocks are time-tracking systems that provide organisms with a survival advantage. Cadmium, one of the toxic metals, is an important occupational and environmental pollutant that damages various organs, especially the kidney. The concluding study proposes that the type of kidney cell and severity of cadmium-induced cellular stress appear to determine the effect of MAPK on cell fate.

Chapter 1 - Intracellular parasites such as *Leishmania, Trypanosoma* or *Toxoplasma* need to maintain a propitious environment within the host cell for its survival. In order to accomplish this goal, they have to manipulate several host's signaling pathways in order to inhibit or diminish harmful actions. The modulation of host responses by the MAPK pathways plays an important role on various stages of parasite entry, maturation, survival and duplication. MAPK pathway consists of the extracellular signal-related kinase (ERK), c-Jun N–terminal kinase or stress-activated protein kinase (JNK or SAPK) and MAPK14. Six major families of MAPK have been identified in mammalian cells-ERK1/2, ERK3/4, ERK5, ERK 7/8, JNK 1/2//3 and the p38 isoforms a/b/c (ERK6/d). Different stimuli such as Gi-coupled receptors (GPCR), growth factors, inflammatory cytokines or a wide range of environmental, oxidative and genotoxic stresses can activate p38 and JNK pathways. The p38 pathway plays an important role in the regulation of apoptosis, cell cycle arrest, growth inhibition, and differentiation JNK has been implicated in a variety of cellular processes including cell proliferation, apoptosis, autophagy, motility, metabolism, and DNA repair. In this review the authors will summarize the present knowledge about MAPK, with special emphasis on p38 and JNK. The main focus will be directed towards how these signaling pathways are engaged during some infections with intracellular parasites and how they are manipulated by the parasites to allow their survival inside host cells.

Chapter 2 - Circadian clocks are intrinsic, time-tracking systems that endow organisms with a survival advantage. At the molecular level, they can be divided into three conceptual components. The first is the pacemaker, dedicated to generating and sustaining circadian rhythms in physiology by receiving and integrating signals from external time cues. The second component is the input, which refers to the pathway through which these cues are perceived and act upon the circadian pacemaker. The third element relates to how the clock affects physiology, which is achieved through the output pathways. There are three major MAPKs: c-JUN N-terminal kinase (JNK), p38, and extracellular signal-regulated kinase (ERK). Varieties of studies have revealed the critical roles of these MAPKs in regulating the pacemaker, input, and output pathways of circadian clocks. The authors describe selected regulatory aspects of circadian clocks in vertebrates, providing an intriguing link between the MAPKs and circadian clocks.

Chapter 3 - Cadmium, one of the toxic metals, is an important occupational and environmental pollutant that damages various organs, especially the kidney. Animal and cultured cell studies show that cadmium exposure induces apoptotic cell death in proximal tubular epithelial cells and glomerular mesangial cells. However, cadmium-induced cellular stress can activate the signaling pathways responsible for both apoptosis and anti-apoptosis. Accumulating evidence indicates that mitogen-activated protein kinases (MAPKs), including extracellular signal-regulated kinase 1/2 (ERK1/2), ERK5 (also known as big MAPK 1 [BMK1]), c-Jun NH_2-terminal kinase (JNK), and p38, are phosphorylated and activated in kidney cells following exposure to cadmium. This short review summarizes the toxicological significance of MAPK activation in cadmium-induced cell death and survival in five types of kidney cells: renal tubular epithelial cells, glomerular mesangial cells, glomerular endothelial cells, podocyte cells, and human embryonic kidney cells. Most studies in kidney cells show that, generally, the activation of the JNK and p38 pathways leads to cell death while the activation of the ERK pathway leads to cell survival. However, the type of kidney cell and severity of cadmium-induced cellular stress appear to determine the effect of MAPKs on cell fate.

In: Mitogen-Activated Protein Kinases ... ISBN: 978-1-53616-138-0
Editor: Charles K. Hester © 2019 Nova Science Publishers, Inc.

Chapter 1

P38 AND JNK AS TARGETS OF PROTOZOAN PARASITES TO MANIPULATE HOST IMMUNE RESPONSE AND SURVIVE INSIDE HOST CELLS

Laila Gutiérrez-Kobeh[1,*], Jorge Rodríguez-González[1,2], Rosalino Vázquez-López[3] and Arturo A. Wilkins-Rodríguez[1]

[1]Unidad de Investigación UNAM-INC, División de Investigación, Facultad de Medicina, UNAM. Mexico City, Mexico
[2]Posgrado en Ciencias Biológicas, Facultad de Medicina, Unidad de Posgrado, UNAM
Ciudad Universitaria, Ciudad de México, México
[3]Microbiology Department, CICSA, Health Science Faculty, Universidad Anáhuac México,
Norte, Huixquilucan, México

---

[*] Corresponding Author's E-mail: lgutierr@unam.mx.

Abstract

Intracellular parasites such as *Leishmania, Trypanosoma* or *Toxoplasma* need to maintain a propitious environment within the host cell for its survival. In order to accomplish this goal, they have to manipulate several host's signaling pathways in order to inhibit or diminish harmful actions. The modulation of host responses by the MAPK pathways plays an important role on various stages of parasite entry, maturation, survival and duplication. MAPK pathway consists of the extracellular signal-related kinase (ERK), c-Jun N–terminal kinase or stress-activated protein kinase (JNK or SAPK) and MAPK14. Six major families of MAPK have been identified in mammalian cells-ERK1/2, ERK3/4, ERK5, ERK 7/8, JNK 1/2//3 and the p38 isoforms a/b/c (ERK6/d). Different stimuli such as Gi-coupled receptors (GPCR), growth factors, inflammatory cytokines or a wide range of environmental, oxidative and genotoxic stresses can activate p38 and JNK pathways. The p38 pathway plays an important role in the regulation of apoptosis, cell cycle arrest, growth inhibition, and differentiation JNK has been implicated in a variety of cellular processes including cell proliferation, apoptosis, autophagy, motility, metabolism, and DNA repair. In this review we will summarize the present knowledge about MAPK, with special emphasis on p38 and JNK. The main focus will be directed towards how these signaling pathways are engaged during some infections with intracellular parasites and how they are manipulated by the parasites to allow their survival inside host cells.

1. Generalities of MAPK

Cells need to establish communication between the external environment and the internal milieu. To achieve this goal, they use a complex array of molecular networks that respond to different stimuli that result in several effects such as changes in metabolic rates, growth, proliferation, differentiation and cell death, among others. Within this complex molecular network, mitogen-activated protein kinases (MAPK) are key participants. MAPKs are a highly conserved family of serine/threonine protein kinases constituted by four members: extracellular signal-regulated kinase 1 and 2 (Erk1/2 or p44/42), the c-Jun N-terminal kinases 1-3 (JNK1-3)/stress activated protein kinases (SAPK1A, 1B, 1C), the p38 isoforms (p38α, β, γ, and δ), and Erk5 (Cargnello and Roux, 2011; Plotnikov et al.,

2011). MAPK signaling pathways can be activated by diverse stimuli such as Gi-coupled receptors (GPCR), growth factors, inflammatory cytokines, mitogens, or a wide variety of environmental, oxidative, and genotoxic stresses (Sui et al., 2014). Such stimuli can activate MAPK cascade in a receptor dependent of independent manner (Cargnello and Roux, 2011; Plotnikov et al., 2011). MAPK signaling cascades involve four sequential activation steps: 1) stimulus reception, 2) activation (phosphorylation) of one or more MAPKK kinases (MAPKKK, MAP3K or MAP3 kinases), 3) MAPKKK activation of MAPK kinase (MAPKK, MAP2K or MAP2 kinase) and 4) activation of MAPK (Jonak et al., 2002; MAPK Group et al., 2002; Tena et al., 2001). MAPKs are activated upon dual phosphorylation of threonine and tyrosine residues in a conserved Thr–X–Tyr motif (X is any amino-acid) (Cuenda and Rousseau, 2007; Morrison, 2012). In order to achieve their biologic effect, MAPK translocate to the nucleus to regulate the transcription and modulate different forms of gene expression (Cargnello and Roux, 2011; Plotnikov et al., 2011b). Several studies have demonstrated that intracellular parasites can modulate the activation/inhibition of MAPK to survive inside host cells (Sarkar et al., 2013) and the MAPKs that are mainly involved in this modulation are p38 and JNK (Vázquez-López et al., 2015).

1.1. p38 Family

p38 was first identified in 1994 by Lee and colleagues as a tyrosine phosphorylated protein in LPS-stimulated macrophages (Lee et al., 1994). The p38 MAPK family is activated upon dual phosphorylation of threonine and tyrosine residues in a conserved Thr-Gly-Tyr motif (Cuenda and Rousseau, 2007; Morrison, 2012) and is composed of four members: p38α, p38β, p38γ (SAPK3 or ERK6), and p38δ (SAPK4) (Cuenda and Rousseau, 2007; Goedert et al., 1997; Jiang et al., 1997; Lechner et al., 1996; Mertens et al., 1996). Isoforms share a 12-amino acid activation loop and differ in affinity in terms of the activating protein, tissue expression, and downstream effect. They participate in the regulation of certain growth factors, kinases, and phosphatases, as well as in the regulation of activating transcription factor 2 (ATF-2), myocyte enhancer factor (MEF2), MAPK activated protein kinase (MAPKAPK), cell division cycle 25 (CDC25), or mitogen- and stress-activated protein kinase 1 and 2 (MSK1/2). Their activation triggers cellular survival, development, and maturation (Bulavin et al., 2001; Chang and Karin, 2001; Davies et al., 2000; Plotnikov et al., 2011b; Tibbles and Woodgett, 1999; Waskiewicz and Cooper, 1995).

The p38α isoform, commonly referred to as p38, as well as the p38β isoform, are present in almost every tissue. Contrary to that, p38γ and p38δ isoforms have a more restricted localization; the former being present in skeletal muscle and the latter in lungs, kidneys, testicles, pancreas, and small intestine (Cuenda and Rousseau, 2007; Ono and Han, 2000). The activation of p38 starts when stress conditions, such as genotoxic or osmotic shock, activate mitogen-activated protein kinase kinase kinase 3 (MEKK3), MEK4 or TAK1, which are phosphorylated downstream into MKK3, MKK6, and very rarely MKK4, which in turn activate p38 by phosphorylating specifically at Thr180 and Tyr182 sites. This phosphorylation process produces conformational changes that lead to the enzyme binding with ATP and the acceptor substrate of the phosphate (Gao et al., 2012; Geng et al., 2017; Mihaly et al., 2014; Morioka et al., 2014; Sassmann-Schweda et al., 2016)

1.2. Jun N-Terminal Kinase (JNK) Family

JNK proteins are also known as stress-associated MAPKs or stress-activated protein kinases (SAPKS). They participate in cellular growth, differentiation, and apoptosis (Khan et al., 2001; K.-W. Lee et al., 2012) as a response to several stress signals, such as hyperosmolarity, UVR or gamma radiation, ischemic damage, thermal shock, toxins, peroxides, protein synthesis inhibitors (anisomycin), anti-neoplasic drugs, and inflammatory cytokines, among others (Khan et al., 2001). Stress signals activate TAK1, MAP3K, ASK1, and ASK2, which in turn activate MEK4 and MEK7 through phosphorylation of two specific serine (Ser) and Thr residues.

MEK4 and MEK7, also known as MKK4 (SEK1/JNKK1) or MKK7 (SEK2/JNKK2) are both MAPKK, and phosphorylate JNK in Thr-Pro-Tyr specific residues (Geng et al., 2017; Ip and Davis, 1998; Khan et al., 2001; Mihaly et al., 2014; Morioka et al., 2014; Rubinfeld and Seger, 2005; Sassmann-Schweda et al., 2016) JNK are codified by three genes: NK1 (46 kDa) (Dérijard et al., 1994; Kyriakis et al., 1994), JNK2 (55 kDa) (Kallunki et al., 1994) and JNK3 (48 kDa) (Gupta et al., 1996). These genes are subject to at least ten types of alternative splicing in order to generate the different isoforms that until now are the following: JNK1α1, JNK1α2, JNK1β1, JNK1β2, JNK2α1, JNK2α2, JNK2β1, JNK2β2, JNK3α1, and JNK3α2. Although these isoforms are physically different, their biological roles are similar (Dreskin et al., 2001). JNK1 and JNK2 isoforms are expressed in all tissues, while JNK3 isoform is found predominantly in nervous tissue, and to a lesser extent in the heart and sperm (Chang and Karin, 2001; Chen et al., 2009; Kumagae et al., 1999).

Although JNK1, JNK2, and JNK3 can all induce apoptosis, evidence suggests that each protein induces apoptosis through a different pathway. It has been demonstrated that all of them associate with p53, to activate proapoptotic gene expression, such as Bax or Puma, but interestingly, their expression varies with respect to p53. In the case of JNK1, its expression is inversely proportional to p53, as opposed to JNK2 expression, which is directly proportional to p53. Both JNK2 and JNK3 can phosphorylate p53, while JNK1 can only modify it post-transcriptionally (Hu, et al., 1997; Tafolla et al., 2005).

As already mentioned, MAPKs are key participants in the development of apoptosis. Intracellular parasites need to inhibit apoptosis of host cells in order to survive inside them. For this reason, MAPKs represent a crucial target for intracellular parasites to modulate host cell apoptosis and survive inside cells. Different intracellular parasites have been shown to modulate MAPKs. Examples of parasites that perform this modulation are *Leishmania*, *Trypanosoma*, *Toxoplasma*, *Plasmodium*, whose strategies to modulate MAPKs will be explained below.

2. MODULATION OF MAPKS
BY INTRACELLULAR PARASITES

2.1. *Leishmania*

Leishmania species are obligate intracellular parasites that cause a spectrum of human diseases from self-healing cutaneous lesions to a fatal visceral form of the disease. It is estimated that there are 12 million people infected with this parasite in 88 countries (Hotez, et al., 2012). Leishmania parasites are transmitted by sand flies in the form of flagellated promastigotes to the mammalian host where they are phagocytosed primarily by macrophages (Mφ) and dendritic cells (DC). After internalization, promastigotes differentiate to small, non-motile amastigotes that proliferate within the host cell phagolysosome (Kane and Mosser, 2000). Internalization into host cells permits intracellular microorganisms to avoid destruction by the immune system; nevertheless, cells can counteract infection by initiating their own death by apoptosis. *Leishmania* has the capacity to modulate diverse processes of host cells and one very important mechanism is through MAPKs phosphorylation since these kinases are involved in several cells processes such as growth, proliferation, differentiation, cytokine production and cell death.

Different studies have shown that *Leishmania* has the capacity to both activate and inhibit different MAPKs. In terms of activation, it has been shown that the transformation of human monocytes of the cell line U-937 to macrophages activates ERK1/2 signaling pathway. Interestingly, when cells are infected with recently-isolated *L. major* amastigotes, there is a reduction of ERK 1/2 phosphorylation induced by PMA. Also, it was observed that the inhibition of protein tyrosine kinases of infected cells reversed the reduction in ERK1/2 phosphorylation. This suggests that the parasite needs the activation of tyrosine kinases to modulate ERK phosphorylation (Guizani-Tabbane et al., 2000).

Another crucial kinase that is modulated by *Leishmania* is Akt. It has been shown that the infection of bone marrow-derived macrophages (BMDM) of BALB/c mice with *L. amazonensis* promastigotes results in the sustained phosphorylation of Akt. The use of a PI3K inhibitor reverses Akt phosphorylation present in *Leishmania*-infected BMDM, which indicates that the activity of PI3K is needed for the infection-induced Akt phosphorylation. Moreover, PI3K inhibition results in IL-12 production, which is why the activation of this kinase by the parasite is crucial for the inhibition of IL-12. It is outstanding that the inhibition of Akt induces a major production of IL-12 in comparison to the sole inhibition of PI3K, which could be due to the initial interaction of the parasite with molecules of the cell surface which is sufficient to activate PI3K and inhibit IL-12 production. On the other hand, internalized parasites manage to block more effectively PI3K/Akt signaling pathway. It will be very interesting to analyze the possibility that AKT inhibition blocks a key point in the amplification of the signaling cascade that is causing the steep increase in IL-12 production. Akt could have a central role in signal transduction due to the fact that the inhibition of this kinase in *Leishmania*-infected drastically diminishes ERK1/2 phosphorylation and increases p38 phosphorylation. Contrarily, in infected cells in which Akt has not been inhibited, there is a higher ERK1/2 phosphorylation and lower p38 phosphorylation. Additionally, Akt inhibition in infected cells is responsible for the restoration of IL-12 production, not so the inhibition of ERK1/2 o p38. The fact that Akt inhibition is enough for restoring IL-12 production is striking and in this respect authors point out that it is possible that the cell manages to initiate signaling pathways needed for IL-12 production. Nevertheless, *Leishmania* is capable of sensing the activation of these routes and in response modulates signaling cascades such as PI3K/Akt, necessary to counteract host cell response (Ruhland and Kima, 2009).

In a different study performed with bone marrow-derived dendritic cells (BMDC) from CH3HeB/FeJ mice infected with promastigotes or amastigotes of *L. major* and *L. amazonensis* aimed at determining the participation of ERK1/2 signaling pathway in BMDC maturation and IL-12 production. It was shown that the infection of cells with

L. amazonensis amastigotes induced a marked increase in ERK1/2 phosphorylation in times shorter that 1 h, but not with the infection with *L. major* amastigotes. Nevertheless, when the infection was initiated with promastigotes of either species, they did not observe such a marked increase in ERK1/2 phosphorylation compared with the infection *L. amazonensis* amastigote. This would indicate that *L. amazonensis* amastigotes have developed a strategy to specifically activate ERK1/2. With the aim of contributing with more arguments in favor of this hypothesis, it will be important to increase the time of study to 24 h, due to the fact that according to what has been reported by Ruhland and Kima, the infection initiated with *L. amazonensis* promastigotes manages to phosphorylate ERK1/2 in a sustained manner for up to 24 h. Interestingly, they found that the specific inhibition of ERK1/2 increases CD40 along with IL-12 production. In this study it was also found that the infection of cells with *L. amazonensis* or *L. major* amastigotes induces p38 and JNK phosphorylation in times shorter than 1 h, although the difference regarding the respect to the basal condition was not statistically significant. Nevertheless, it will be interesting to find out if the same occurs in times post-infection higher to 1 h (Boggiatto et al., 2014).

In another study performed with BMDM from C57BL/6 and infected with *L. mexicana* promastigotes or amastigotes deficient in the production of cysteine peptidase B (CBP), it was observed that both morphological stages induce ERK1/2, JNK1/2, and p38 phosphorylation. Additionally, it was observed that the infection of BMDM derived from mice deficient in TLR4, completely abrogated MAPKs activation. Authors point out that MAPKs activation is independent of the parasite morphological stage used for the initiation of infection, since the phosphorylation levels between MAPKs are very similar, no matter if cells were infected with CBP-deficient promastigotes or amastigotes (Shweash et al., 2011).

MAPK activation has also been observed with *Leishmania* infection along with other effects such as the production of reactive oxygen species (ROS). It has been shown that the infection of peritoneal macrophages from C57BL/6 mice with *L. major* promastigotes JNK phosphorylation along with ROS production. Moreover, it was observed that the use of anti-oxidants diminished JNK phosphorylation in infected cells, which suggested that ROS production is necessary to achieve this effect. Interestingly, it was observed that the incubation of cells with the anti-oxidants or with a JNK specific inhibitor, diminished parasite loads, which could suggest that ROS production along with JNK activation favors *Leishmania* development (Filardy et al., 2014).

Other studies have shown that the treatment with LPS of dendritic cells of the cell line DC2.4 before or after the infection with *L. mexicana* promastigotes diminishes ERK1/2 and p38 phosphorylation, but not of JNK. This suggests that MAPK ERK1/2 and p38 are more related to the modulation of molecules that favor parasite survival such as an increase in IL-10, decrease in IL-12 and DC maturation (Contreras et al., 2014). In regards to the modulation of MAPKs in favor of *Leishmania* survival inside host cells, it has been shown the interaction of neutrophils with *L. major* transiently activates ERK1/2 phosphorylation that results in a delay in the induction of apoptosis, which is reversed with the pharmacological inhibition of ERK1/2 phosphorylation. Additionally, it was shown that the infection leads to the enhanced and sustainable expression of anti-apoptotic proteins such as Bcl-2 and Bfl-1. As downstream events, the release of cytochrome *c* from mitochondria and processing of caspase-6 were inhibited. It was also confirmed that infection with *L. major* results in reduced FAS expression on the surface of neutrophils. The presented data indicate that infection with *L. major* affects both intrinsic as well as extrinsic pathways of neutrophil apoptosis. Enhanced life span of host neutrophils enables the parasite to survive within neutrophils (Sarkar et al., 2013).

Our group has worked for the last few years in the involvement of MAPKs in the survival of *L. mexicana* inside dendritic cells. We have shown that the infection of monocyte-derived dendritic cells (moDC) with *L. mexicana* amastigotes and promastigotes induces a sustained phosphorylation of JNK and p38 starting from 1 h post-infection and up to 24 h. Interestingly, the induction of apoptosis with camptothecin (CPT) in uninfected moDC, also induces the sustained phosphorylation of JNK and p38 up to 24 h. Nevertheless, in CPT-treated moDC and infected with *L. mexicana* promastigotes or amastigotes, there is a significant decrease in

JNK and p38 phosphorylation (Rodríguez-González et al., 2016; Vázquez-López et al., 2015). We also found that the infection of moDC with *L. mexicana* amastigotes increased the phosphorylation of PI3K/Akt, which promotes DC survival (Vázquez-López et al., 2015).

In human monocyte-derived macrophages (hMDM) or derived from THP-1 cells and e infected with *L. donovani* promastigotes it has been shown that in infected cells and treated with sphigosine-1-phosphate (S1P), a lipid whose synthesis is catalyzed by sphingosine kinase 1 (SphK1), reverses ERK1/2 phosphorylation and induces p38 phosphorylation, which has been associated with a decrease in IL-10 mRNA and an increase in IL-12 mRNA. Strikingly, in the presence of a Sphk1 inhibitor, the opposite takes place, ERK1/2 phosphorylation increases and p38 phosphorylation decreases along with IL-10 increase and IL-12 decrease. Coupled with this, it was observed that the infection of cells diminishes SphK1 phosphorylation, which could finally diminish S1P synthesis (Arish et al., 2018).

In general terms, we can conclude that *Leishmania* manages to modulate the phosphorylation of several kinases such as MAPK ERK1/2, JNK1/2, p38 that importantly participate in apoptosis and in PI3K/Akt that play a crucial role in cells survival. While ERK1/2 phosphorylation has been widely related to IL-10 production and parasite survival, p38 phosphorylation has been associated to IL-12 production and a decrease in parasite survival. Nevertheless, it has not been well established to which *Leishmania* survival mechanisms JNK phosphorylation is related to, although it has been shown that a decrease in the activation of JNK inhibits induced apoptosis in host cells. On the other hand, it seems that the activation of PI3K/AKT is highly related to the modulation of ERK1/2 and p38 in order to increase parasite survival in host cells. This strategy that has been developed by the parasite seems to be highly conserved due to the fact that in general terms is independent of the species of *Leishmania* and the genetic background of host cells.

2.2. *Trypanosoma cruzi*

Trypanosoma cruzi is the causal agent of Chagas' disease that affects near 8 million people in Latin America [World Health Organization. Chagas disease (American trypanosomiasis). Fact sheet No. 2015:340]. This parasite has the capacity to infect virtually any cell. Upon infection, parasites reach the cytoplasm where they replicate, lyse the cell and spread to other cells. One of the most important cells for *T. cruzi* duplication are macrophages. For *T. cruzi* to invade and persist in host cells, it has to overcome several defense mechanisms that involve different transduction pathways such as MAPKs. The mechanisms by which *T. cruzi* evades the host's immune response by acting on the MAPKs pathway have been poorly analyzed. However, it has been well established that upon cell invasion, *T. cruzi* subverts signaling pathways. It has been shown that *T. cruzi* is able to induce ERK1/2, but not p38 MAPK activation in macrophages and dendritic cells (Mukherjee et al., 2004) and increase IL-10 with the concomitant decrease in IL-12 production (Poncini et al., 2008). In addition to subverting signaling pathways to evade host defense mechanisms, *T. cruzi* uses host molecules to favor its entry and survival inside host cells. As an example, it has been shown that *T. cruzi* extracellular amastigotes (EAs) enter into HeLa cells through recruitment of both host protein kinase D1 (PKD1) and cortactin to induce PKD1 autophosphorylation and cortactin activation by ERK, which in turn recruits host actin (Bonfim-Melo et al., 2015). These effects intervene with the correct deviation of the immune response to an inflammatory Th1 and allows parasite evasion of the host immune response (Alba Soto et al., 2003, 2010). Also, *T. cruzi* secretes some molecules that activate Toll-like receptors (TLRs), such as TLR2, TLR4, or TLR9, in dendritic cells and macrophages (Tarleton, 2007), which leads to the activation of p38 MAPK and the production of IL-12, favoring an inflammatory response (Ropert et al., 2001; Terrazas et al., 2011). The exposure of dendritic cells to *T. cruzi* antigens (TcAgs) and TLR ligands induced p38 phosphorylation that was dependent on TcAg-macrophage migration inhibitory factor (MIF) synergism. This increased IL-12 production and thus promoted a Th1-type response (Terrazas et al., 2011).

Although several *T. cruzi* molecules induce a proinflammatory response, others have the capacity to hijack host signaling pathways in order to subvert host's immune response (Ouaissi et al., 1995; Hovsepian et al., 2011; Castillo et al., 2013; Ruiz Díaz et al., 2015) through events probably mediated by MAPKs. Among the molecules released by *T. cruzi* that interfere with the host immune response is a 52 kDa protein named Tc52A, that is composed of two homologous domains that share homology with glutathione *S*-transferases (Schöneck et al., 1994). Tc52 exerts an immunomodulatory role through the suppression of T cell proliferation and modulation of IL-12 and IL-10 production and virulence roles (Ouaissi et al., 2002b; Ouaissi and Ouaissi, 2005). In addition to Tc52, other *T. cruzi* proteins have the capacity to modulate host immune response among which glycosylphosphatidylinositol (GPI)-anchored mucins and *trans*-sialidases (TS) play an important role. It has been shown that a GPI-anchored mucin of 40–50 kDa named AgC10 inhibits p38 activation affecting TNF and IL-12 secretion and thus the development of a Th1 response (De Diego et al., 1997; Alcaide and Fresno, 2004). Contrarily, *T. cruzi* GPI-molecules and mucins can also activate MAPKs, as is the case of the activation of ERK1/2 in macrophages with the concomitant decrease in IL-12 production (Ropert et al., (2001). Interestingly, the mucins can also activate p38 at later times than ERK1/2, thus increasing IL-12 synthesis and promoting a Th1 response (Ropert et al., 2001). ERK1/2 can also be activated by *T. cruzi* TS (Chuenkova and Pereira, 2001). Also, TS can stimulate the secretion of IL-10, which deviates the immune response to a Th2 (Ruiz Díaz et al., (2015). Many of the mechanisms employed by *T. cruzi* to subvert host immune response are not fully understood and remain to be elucidated.

As in the case of *Leishmania*, *T. cruzi* has the capacity to modulate MAPKs in its behalf. Nevertheless, the modulation of MAPKs by *T. cruzi* has not been analyzed in the context of host apoptosis inhibition as has been shown for *Leishmania*.

2.3. *Toxoplasma gondii*

Toxoplasma gondii is an obligate intracellular parasite capable of infecting almost all types of nucleated cells and represents an important cause of ocular disease in immunosuppressed and immunocompetent individuals. The infection of host cells by *T. gondii* occurs by gliding motility based on penetration of the host cell plasma membrane. Once inside its host, *T. gondii* resides within a specialized compartment called a parasitophorous vacuole, which resists phagolysosomal fusion and actively recruits host cell organelles. *T. gondii* has developed multiple mechanisms to avoid immune elimination and has an outstanding capacity of hijacking host cell apoptotic machinery to ensure its survival. Interestingly, it can induce either an anti- or pro-apoptotic response depending on several factors such as the parasite virulence and load, as well as the host cell type (Contreras-Ochoa et al., 2013). In regards to a pro-apoptotic response, it has been shown that T. *gondii* has the capacity to induce apoptosis in several cell types such as spleen cells (Gravilescu and Denkers, 2001), neurons (El-Sagaff et al., 2005), and choriocarcinoma cells (Angeloni et al., 2013). One of the mechanisms that has shown to be involved in *T. gondii* induction of apoptosis is endoplasmic reticulum stress (ERS) (Wang et al., 2014; Zhou et al., 2015) and probably rhoptry proteins are able to trigger ERS-mediated apoptosis (Chang 2015).

On the other hand, *T. gondii* infection can also interfere with the apoptotic process in host cells, thereby enhancing its survival (Nash et al., 1998; Luder and Gross 2005; Carmen et al., 2006; Luder et al., 2009). It has been reported that this parasite can modulate multiple signaling pathways in their host cells in order to inhibit apoptosis, ensuring in this way its survival and persistence during infection. One of the mechanisms employed by *T. gondii* to inhibit host cell apoptosis is the inactivation of caspases, in particular, caspase 3, 8, and 9 (Carmen et al., 2006; Kim et al., 2006).

MAPKs have been involved in the suppression of host cell apoptosis by *T. gondii*. As in the case of *Leishmania*, different reports have demonstrated activation, but also inhibition of MAPKs. It has been shown that *T. gondii* has the capacity to rapidly activate all three MAPKs signaling modules upon infection of macrophages and phosphorylation returns to basal levels within 2 h post-infection (Valere et al., 2003). Contrarily, it has also been shown that the pre-infection with *T. gondii* suppressed the phosphorylation of p38 MAPK, but not ERK1/2 with the concomitant decrease in the expression of Bad and Bax and increase in the expression of Bcl-2, which resulted in the inhibition of apoptosis in retinal pigment epithelial cells. It was shown that the p38 MAPK downstream molecule HO-1 was partly involved in H_2O_2-induced apoptosis in ARPE-19 cells (Choi et al., 2011).

Recently it has been shown that exosomes secreted by *T. gondii* were able to activate JNK and increase the production of IL-12, TNF-α, and IFN-γ in macrophages of the cell line RAW264.7 (Li et al., 2018).

2.4. *Plasmodium* spp.

The etiologic agents of malaria are protozoan parasites in the genus *Plasmodium* and are responsible for 216 million new cases and 445,000 deaths worldwide in 2016 (WHO, 2016). In the mammalian host, *Plasmodium* parasites infect primarily hepatocytes and erythrocytes, and modulation of apoptosis by this parasite in both host cell types has been found to be crucial during infection. After transmission by the *Anopheles* mosquito, *Plasmodium* sporozoites are rapidly transported to the liver, where they invade and develop within hepatocytes before reaching erythrocytes (Prudencio et al., 2006). In the liver, sporozoites transmigrate through the cytosol of multiple hepatocytes, causing wounding in the traversed cells with the release of the hepatocyte growth factor (HGF), which helps the parasite to reach a final hepatocyte in which it will reside and multiply (Carrolo et al., 2003).

MAPKs activation has been shown to participate at different levels of *Plasmodium* infection. Starting with the interaction of the parasites with *Anopheles* mosquitoes, JNK signaling has been implicated in mosquito defense against malaria parasites. It has been shown that human insulin-like growth factor 1 (IGF1) induces NOS expression to mediate inhibition of *P. falciparum* development via enhanced JNK activation in the midgut epithelium of *A. stephensi* (Drexler et al., 2014). Also, in *A. gambiae*, JNK activation has been linked to the regulation of heme peroxidase 2 and NAPDH oxidase 5, both of which function with NOS to opsonize parasites, leading to TEP1-mediated complement-like elimination in the mosquito host (Oliveira et al., 2012). Other authors have proposed that a protein expressed in the ookinete stage of *P. falciparum*, *Pfs*47, disrupts *A. gambiae* JNK signaling to facilitate parasite evasion of the immune response and enhance parasite survival in the mosquito (Ramphul et al., 2015). In another work, an *A. gambiae* ortholog of a MAP phosphatase (MKP), was silenced with iRNA and increased resistance in the mosquito against infection with *Plasmodium berghei* (Garver et al., 2013).

CONCLUSION

Intracellular parasites are outstanding in their capacity to manipulate host cells in order to evade the immune response and maintain themselves inside cells. This great endeavor is achieved through mechanisms that differ among different parasites, stages, life cycles, host cells, etc.. A strategy that is common among different parasites and shared with other microorganisms is the capacity to use host's molecules as targets to evade the immune response. MAPKs are key participants in the immune response through the signaling cascades that they unchain and that result in diverse effects such as growth, proliferation, cytokine production, differentiation and cell death, among others. These kinases have been shown to be modulated by different intracellular parasites such as *Leishmania, Trypanosoma, Toxoplasma* and *Plasmodium*, to cite some examples, and this regulation results in many cases beneficial to the parasite in order to persist in the host. In this chapter we have summarized very interesting work that has been published in this regard. Many issues about the manipulation of host cells by intracellular parasites remain to be elucidated and represent a transcendental line of investigation due to the fact that these signaling cascades that are modulated by pathogens represent possible targets for the treatment and prevention of the diseases caused by them.

REFERENCES

Alba Soto, C. D., Mirkin, G. A., Solana, M. E., and González Cappa, S. M. (2003). Trypanosoma cruzi infection modulates in vivo expression of major histocompatibility complex class II molecules on antigen-presenting cells and T-cell stimulatory activity of dendritic cells in a strain-dependent manner. *Infection and Immunity.* 71, 1194–1199. doi: 10.1128/IAI.71.3.1194-1199.2003.

Alba Soto, C. D., Solana, M. E., Poncini, C. V., Pino-Martinez, A. M., Tekiel, V., and González-Cappa, S. M. (2010). Dendritic cells devoid of IL-10 induce protective immunity against the protozoan parasite *Trypanosoma cruzi. Vaccine* 28, 7407–7413. doi: 10.1016/j.vaccine. 2010.08.105.

Alcaide, P. and Fresno, M. (2004). AgC10, a mucin from *Trypanosoma cruzi*, destabilizes TNF and cyclooxygenase-2 mRNA by inhibiting mitogen-activated protein kinase p38. *European Journal of Immunology,* 34, 1695–1704. doi: 10.1002/eji.200324660.

Angeloni, M. B., Guirelli, P. M., Franco, P. S., Barbosa, B. F., Gomes, A. O., Castro, A. S., Silva, N. N., Martins-Filho, O. A., Mineo, T. W., Silva, D. A., Mineo, J. R., and Ferro, E. A. (2013). Differential apoptosis in BeWo cells after infection with highly (RH) or moderately (ME49) virulent strains of *Toxoplasma gondii* is related to the cytokine profile secreted, the death receptor Fas expression and phosphorylated ERK1/2 expression. *Placenta*, 34(11), 973–982. doi: 10.1016/j.placenta. 2013.09.005.

Arish, M., Husein, A., Ali, R., Tabrez, S., Naz, F., Ahmad, M. Z. and Rub, A. (2018). Sphingosine-1-phosphate signaling in *Leishmania donovani* infection in macrophages. *PLOS Neglected Tropical Diseases*, 12(8), e0006647. doi: 10.1371/journal.pntd.0006647.

Boggiatto, P. M., Martinez, P. A., Pullikuth, A., Jones, D. E., Bellaire, B., Catling, A. and Petersen, C. (2014) Targeted extracellular signal-regulated kinase activation mediated by *Leishmania amazonensis* requires MP1 scaffold. *Microbes and Infection*, 16(4), 328–336. doi: 10.1016/j.micinf.2013.12.006.

Bonfim-Melo, A., Zanetti, B. F., Ferreira, É. R., Vandoninck, S., Han, S. W., Van Lint, J., Mortara, L. A., and Bahia, D. et al., (2015). *Trypanosoma cruzi* extracellular amastigotes trigger the protein kinase D1-cortactin-actin pathway during cell invasion. *Cellular Microbiology* 17, 1797–1810. doi: 10.1111/cmi.12472.

Bulavin, D. V., Higashimoto, Y., Popoff, I. J., Gaarde, W. A., Basrur, V., Potapova, O., Appella, E. and Fornace, A. J. Jr. (2001). Initiation of a G2/M checkpoint after ultraviolet radiation requires p38 kinase. *Nature*, 411(6833), 102–107. doi.org/10.1038/35075107.

Cargnello, M. and Roux P. P. (2011). Activation and function of the MAPKs and their substrates, the MAPK-activated protein kinases. *Microbiology and Molecular Biology Reviews*, 75(1), 50–83. doi: 10.1128/MMBR.00031-10.

Carmen, J. C., Hardi, L., and Sinai, A. P. (2006) *Toxoplasma gondii* inhibits ultraviolet light-induced apoptosis through multiple interactions with the mitochondrion-dependent programmed cell death pathway. *Cellular Microbiology*, 8 (2), 301–315. doi: 10.1111/j.1462-5822.2005.00622.x.

Carrolo, M., Giordano, S., Cabrita-Santos, L., Corso, S., Vigário, A. M., Silva, S., Leirião, P., Carapau, D., Armas-Portela, R., Comoglio, P. M., Rodriguez, A., and Mota, M. M. (2003). Hepatocyte growth factor and its receptor are required for malaria infection. *Nature Medicine*, 9(11):1363-1369. doi: 10.1038/nm947.

Castillo, C., Villarroel, A., Duaso, J., Galanti, N., Cabrera, G., Maya, J. D., and Kemmerling, U. (2013). Phospholipase C gamma and ERK1/2 Mitogen Activated Kinase Pathways are differentially modulated by *Trypanosoma cruzi* during tissue invasion in human placenta. *Experimental Parasitology,* 133, 12–17.doi:10.1016/j.exppara.2012.10.012.

Chang, L. and Karin, M. (2001). Mammalian MAP kinase signaling cascades. *Nature*, 410(6824), 37–40. doi.org/10.1038/35065000.

Chang, S., Shan, X., Li, X., Fan, W., Zhang, S. Q., Zhang, J., Jiang, N., Ma, D., and Mao, Z. (2015). *Toxoplasma gondii* rhoptry protein ROP16 mediates partially SH-SY5Y cells apoptosis and cell cycle arrest by directing ser15/37 phosphorylation of p53. *International Journal of Biological Sciences*, 11(10), 1215–1225. doi: 10.7150/ijbs.10516.

Chen, F., Beezhold, K. and Castranova, V. (2009). JNK1, a potential therapeutic target for hepatocellular carcinoma. *Biochimica et Biophysica Acta*, 1796(2), 242–251. doi: 10.1016/j.bbcan.2009.06.005.

Choi, S. H., Park, S. H., Cha G. H., Quan J. H., Chang, N. S., Ahn, M. H., Shin, D. W., and Lee, Y. H. (2011). *Toxoplasma gondii* protects against H2O2-induced apoptosis in ARPE-19 cells through the transcriptional regulation of apoptotic elements and downregulation of the p38 MAPK pathway. *Acta Ophtalmologica*, 89(4), e350-6. doi: 10.1111/j.1755-3768.2011.02113.x.

Chuenkova, M. V. and Pereira, M. A. (2001). The *T. cruzi* trans-sialidase induces PC12 cell differentiation via MAPK/ERK pathway. *Neuroreport* 4, 3715–3718. doi: 10.1097/00001756-200112040-00022.

Contreras, I., Estrada, J. A., Guak, H., Martel, C., Borjian, A., Ralph, B., Shio, M. T., Fournier, S., Krawczyk, C. M. and Olivier, M. (2014) Impact of *Leishmania mexicana* infection on dendritic cell signaling and functions. *PLoS Neglected Tropical Diseases*, 8(9), e3202. doi: 10.1371/journal.pntd.0003202.

Contreras-Ochoa C. O., Lagunas-Martínez, A., Belkind-Gerson, J., Díaz-Chávez, J., and Correa, D. (2013). *Toxoplasma gondii* invasion and replication within neonate mouse astrocytes and changes in apoptosis related molecules. *Experimental Parasitology*, 134, 256–265. doi: 10.1016/j.exppara.2013.03.010.

Cuenda, A. and Rousseau, S. (2007). p38 MAP-Kinases pathway regulation, function and role in human diseases. *Biochimica et Biophysica Acta*, 1773(8), 1358–1375. doi.org/10.1016/J.BBAMCR.2007.03.010.

Davies, S. P., Reddy, H., Caivano, M. and Cohen, P. (2000). Specificity and mechanism of action of some commonly used protein kinase inhibitors. *The Biochemical Journal*, 351(Pt 1), 95–105. doi: 10.1042/bj3510095.

de Diego, J., Punzón, C., Duarte, M., and Fresno, M. (1997). Alteration of macrophage function by a *Trypanosoma cruzi* membrane mucin. *Journal of Immunology*, 15, 4983–4989.

Dérijard, B., Hibi, M., Wu, I. H., Barrett, T., Su, B., Deng, T., Karin M. and Davis, R. J. (1994). JNK1: a protein kinase stimulated by UV light and Ha-Ras that binds and phosphorylates the c-Jun activation domain. *Cell*, 76(6), 1025–1037. doi: https://doi.org/10.1016/0092-8674(94)90380-8.

Dreskin, S. C., Thomas, G. W., Dale, S. N. and Heasley, L. E. (2001). Isoforms of Jun kinase are differentially expressed and activated in human monocyte/macrophage (THP-1) cells. *Journal of Immunology*, 166(9), 5646–5653. Retrieved from http://www.ncbi.nlm.nih.gov/pubmed/11313405.

Drexler, A. L., Pietri, J. E., Pakpour, N., Hauck, E., Wang, B., Glennon, E. K., Georgis M., Riehle M. A. and Luckhart, S. (2014). Human IGF1 regulates midgut oxidative stress and epithelial homeostasis to balance lifespan and *Plasmodium falciparum* resistance in *Anopheles stephensi*. *PLoS Pathogens*, 10(6): e1004231. doi: 10.1371/journal.ppat.1004231.

El-Sagaff, S., Salem, H. S., Nichols, W., Tonkel, A. K., and Abo-Zenadah, N. Y. (2005). Cell death pattern in cerebellum neurons infected with *Toxoplasma gondii*. *Journal of the Egyptian Society of Parasitology*, 35, 809–18.

Filardy, A. A., Costa-da-Silva A. C., Koeller, C. M., Guimarães-Pinto, K., Ribeiro-Gomes, F. L., Lopes, M. F., Heise, N., Freire-de-Lima, C. G., Nunes, M. P. and DosReis, G. A. (2014). Infection with *Leishmania major* induces a cellular stress response in macrophages. *PLoS One*, 9(1), e85715. doi: 10.1371/journal.pone.0085715.

Gao, W., Sun, W., Qu, B., Cardona, C. J., Powell, K., Wegner, M., Shi, Y. and Xing, Z. (2012). Distinct regulation of host responses by ERK and JNK MAP kinases in swine macrophages infected with pandemic (H1N1) 2009 influenza virus. *PLoS ONE*, 7(1), e30328. doi: 10.1371/journal.pone.0030328.

Garver, L. S., de Almeida Oliveira, G., and Barillas-Mury, C. (2013). The JNK pathway is a key mediator of *Anopheles gambiae* antiplasmodial immunity. *PLOS Pathogens*, 9(9), e1003622. doi:10.1371/journal.ppat.1003622.

Geng, J., Ito, Y., Shi, L., Amin, P., Chu, J., Ouchida, A. T., Mookhtiar A. K., Zhao, H., Xu, D., Shan, B., Najafov, A., Akira, S. and Yuan, J.

(2017). Regulation of RIPK1 activation by TAK1-mediated phosphorylation dictates apoptosis and necroptosis. *Nature Communications*, 8(1), 359. doi: 10.1038/s41467-017-00406-w.

Goedert, M., Cuenda, A., Craxton, M., Jakes, R. and Cohen, P. (1997). Activation of the novel stress-activated protein kinase SAPK4 by cytokines and cellular stresses is mediated by SKK3 (MKK6); comparison of its substrate specificity with that of other SAP kinases. *The EMBO Journal*, 16(12), 3563–3571. doi: 10.1093/emboj/16.12.3563.

Gavrilescu, L. C. and Denkers, E. Y. (2001). IFN-gamma overproduction and high level apoptosis are associated with high but not low virulence *Toxoplasma gondii* infection. *Journal of Immunology*, 167, 902–909. doi: https://doi.org/10.4049/jimmunol.167.2.902.

Guizani-Tabbane, L., Ouni, I., Sassi, A. and et al., (2000) *Leishmania major* induces deactivation of extracellular signal regulated kinases 2 in human U937 macrophage like cells. *Archives de l'Institut Pasteur de Tunis*, 77(1–4), 45–50. Available at: http://www.ncbi.nlm.nih.gov/pubmed/14658227.

Gupta, S., Barrett, T., Whitmarsh, A. J., Cavanagh, J., Sluss, H. K., Dérijard, B. and Davis, R. J. (1996). Selective interaction of JNK protein kinase isoforms with transcription factors. *The EMBO Journal*, 15(11), 2760–2770. Retrieved from http://www.ncbi.nlm.nih.gov/pubmed/8654373.

Hotez, P. J., Savioli, L., and Fenwick, A. (2012). Neglected tropical diseases of the Middle East and North Africa: review of their prevalence, distribution, and opportunities for control. *PLoS Neglected Tropical Diseases* 6, e1475. doi: 10.1371/journal.pntd.0001475.

Hovsepian, E., Mirkin, G. A., Penas, F., Manzano, A., Bartrons, R., and Goren, N. B. (2011). Modulation of inflammatory response and parasitism by 15-Deoxy-Δ12,14 prostaglandin J2 in *Trypanosoma cruzi*-infected cardiomyocytes. *International Journal for Parasitology*, 41, 553–562. doi: 10.1016/j.ijpara.2010.12.002.

Hu, M. C. T., Qiu, W. R., and Wang, Y. P. (1997). JNK1, JNK2 and JNK3 are p53 N-terminal serine 34 kinases. *Oncogene*, 15(19), 2277–2287. doi: 10.1038/sj.onc.1201401.

Ip, Y. T. and Davis, R. J. (1998). Signal transduction by the c-Jun N-terminal kinase (JNK)--from inflammation to development. *Current Opinion in Cell Biology*, 10(2), 205–219. doi: 10.1016/S0955-0674(98)80143-9.

Jiang, Y., Gram, H., Zhao, M., New, L., Gu, J., Feng, L., Padova F. D., Ulevitch R. J., and Han, J. (1997). Characterization of the structure and function of the fourth member of p38 group mitogen-activated protein kinases, p38delta. *The Journal of Biological Chemistry*, 272(48), 30122–30128. http://www.ncbi.nlm.nih.gov/pubmed/9374491.

Jonak, C., Okrész, L., Bögre, L., and Hirt, H. (2002). Complexity, cross talk and integration of plant MAP kinase signalling. *Current Opinion in Plant Biology*, 5(5), 415–424. doi.org/10.1016/S1369-5266(02)00285-6.

Kallunki, T., Su, B., Tsigelny, I., Sluss, H. K., Dérijard, B., Moore, G., Davis, R., and Karin, M. (1994). JNK2 contains a specificity-determining region responsible for efficient c-Jun binding and phosphorylation. *Genes & Development*, 8(24), 2996–3007. http://www.ncbi.nlm.nih.gov/pubmed/8001819.

Kane, M. M. and Mosser, D. M. (2000). *Leishmania* parasites and their ploys to disrupt macrophage activation. *Current Opinion in Hematology* 7, 26-31.

Khan, S., Koepke, A., Jarad, G., Schlessman, K., Cleveland, R. P., Wang, B., Konieczkowski, M., and Schelling, J. R. (2001). Apoptosis and JNK activation are differentially regulated by Fas expression level in renal tubular epithelial cells. *Kidney International*, 60(1), 65–76. doi: 10.1046/j.1523-1755.2001.00771.x.

Kim, J. Y., Ahn, M. H., Jun, H. S., Jung, J. W., Ryu, J. S., Min, D. Y. (2006). *Toxoplasma gondii* inhibits apoptosis in infected cells by caspase inactivation and NF-kappaB activation. *Yonsei Medical Journal*, 47(6):862-9. doi: 10.3349/ymj.2006.47.6.862.

Kumagae, Y., Zhang, Y., Kim, O. J., and Miller, C. A. (1999). Human c-Jun N-terminal kinase expression and activation in the nervous system. *Brain Research. Molecular Brain Research*, 67(1), 10–17. Retrieved from http://www.ncbi.nlm.nih.gov/pubmed/10101227.

Kyriakis, J. M., Banerjee, P., Nikolakaki, E., Dai, T., Rubie, E. A., Ahmad, M. F., Avruch, J., and Woodgett, J. R. (1994). The stress-activated protein kinase subfamily of c-Jun kinases. *Nature*, 369(6476), 156–160.

Lechner, C., Zahalka, M. A., Giot, J. F., Møller, N. P., and Ullrich, A. (1996). ERK6, a mitogen-activated protein kinase involved in C2C12 myoblast differentiation. *Proceedings of the National Academy of Sciences of the United States of America*, 93(9), 4355–4359. http://www.ncbi.nlm.nih.gov/pubmed/8633070.

Lee, J. C., Laydon, J. T., McDonnell, P. C., Gallagher, T. F., Kumar, S., Green, D., McNulty, D., Blumenthal M. J., Heys J. R., Landvatter S. W., Strickler J. E., McLaughlin, M. M., Siemens I. R., Fisher S. M., Livi, G. P., White, J. R., Adams J. L., and Young, P. R. (1994). A protein kinase involved in the regulation of inflammatory cytokine biosynthesis. *Nature*, 372(6508), 739–746. doi:1038/372739a0.

Lee, K. W., Zhao, X., Im, J. Y., Grosso, H., Jang, W. H., Chan, T. W., Sonsalla, P. K., German, D. C., Ichijo, H., Junn, E., and Mouradian, M. M. (2012). Apoptosis Signal-Regulating Kinase 1 Mediates MPTP Toxicity and Regulates Glial Activation. *PLoS One*, 7(1), e29935. doi: 10.1371/journal.pone.0029935.

Li, Y., Xiu, F., Mou, Z., Xue, Z., Du, H., Zhou, C., Li, Y., Shi, Y., He, S., and Zhou, H. (2018). Exosomes derived from *Toxoplasma gondii* stimulate an inflammatory response through JNK signaling pathway. *Nanomedicine*, 13 (4). https://doi.org/10.2217/nnm-2018-0035.

Lüder, C. G. and Gross, U. (2005). Apoptosis and its modulation during infection with *Toxoplasma gondii*: molecular mechanisms and role in pathogenesis. *Current Topics in Microbiology and Immunology*, 289, 219-37.

MAPK Group, M. G. (Kazuya I., Ichimura, K., Shinozaki, K., Tena, G., Sheen, J., Henry, Y., Champion, A., Kreis, M., Zhang, S., Hirt, H., Wilson, C., Heberle-Bors, E., and Walker, J. C. (2002). Mitogen-activated protein kinase cascades in plants: a new nomenclature. *Trends in Plant Science*, 7(7), 301–308. doi.org/10.1016/S1360-1385(02) 02302-6.

Mertens, S., Craxton, M., and Goedert, M. (1996). SAP kinase-3, a new member of the family of mammalian stress-activated protein kinases. *FEBS Letters*, 383(3), 273–276. http://www.ncbi.nlm.nih.gov/pubmed/8925912.

Mihaly, S. R., Ninomiya-Tsuji, J., and Morioka, S. (2014). TAK1 control of cell death. *Cell Death and Differentiation*, 21(11), 1667–1676. doi: 10.1038/cdd.2014.123.

Morioka, S., Broglie, P., Omori, E., Ikeda, Y., Takaesu, G., Matsumoto, K., and Ninomiya-Tsuji, J. (2014). TAK1 kinase switches cell fate from apoptosis to necrosis following TNF stimulation. *The Journal of Cell Biology*, 204(4), 607–623. doi: 10.1083/jcb.201305070.

Morrison, D. K. (2012). MAP kinase pathways. *Cold Spring Harbor Perspectives in Biology*, 4(11) 1-5. doi.org/10.1101/cshperspect.a011254.

Mukherjee, S., Huang, H., Petkova, S. B., Albanese, C., Pestell, R. G., Braunstein, V. L., Christ, G. J., Wittner, M., Lisanti, M. P., Berman, J. W., Weiss, L. M., and Tanowitz, H. B. (2004). *Trypanosoma cruzi* infection activates extracellular signal-regulated kinase in cultured endothelial and smooth muscle cells. *Infection and Immunity*, 72(9), 5274–5282. doi: 10.1128/IAI.72.9.5274-5282.2004.

Nash, P. B., Purner, M. B., Leon, R. P., Clarke, P., Duke R. C., and Curiel, T. J. (1998). *Toxoplasma gondii*-infected cells are resistant to multiple inducers of apoptosis. *Journal of Immunology*, 160, 1824-1830; http://www.jimmunol.org/content/160/4/1824.

Oliveira, G. and Barillas-Mury C. (2013). The JNK pathway is a key mediator of *Anopheles gambiae* antiplasmodial immunity. *PLoS Pathogens*, 9(9): e1003622. doi: 10.1371/journal.ppat.1003622.

Oliveira, G. de A., Lieberman, J., and Barillas-Mury, C. (2012). Epithelial nitration by a peroxidase/NOX5 system mediates mosquito antiplasmodial immunity. *Science*, 335(6070), 856–859. doi: 10.1126/science.1209678.

Ono, K. and Han, J. (2000). The p38 signal transduction pathway: activation and function. *Cellular Signalling*, 12(1), 1–13. doi.org/10.1016/S0898-6568(99)00071-6.

Ouaissi, A., Guevara-Espinoza, A., Chabe, F., Gomez-Corvera, R., and Taibi, A. (1995). A novel and basic mechanism of immunosuppression in Chagas' disease: *Trypanosoma cruzi* releases in vitro and in vivo a protein which induces T cell unresponsiveness through specific interaction with cysteine and glutathione. *Immunology Lett.* 48, 221–224. doi: 10.1016/0165-2478(95)02463-8.

Ouaissi, A. and Ouaissi, M. (2005). Molecular basis of *Trypanosoma cruzi* and *Leishmania interaction* with their host(s): exploitation of immune and defense mechanisms by the parasite leading to persistence and chronicity, features reminiscent of immune system evasion strategies in cancer dis. *Archivum Immunologiae et Therapiae Experimentalis* 53, 102–114.

Ouaissi, A., Ouaissi, M., and Sereno, D. (2002). Glutathione S-transferases and related proteins from pathogenic human parasites behave as immunomodulatory factors. *Immunology Letters* 81, 159–164. doi: 10.1016/S0165-2478(02)00035-4.

Plotnikov, A., Zehorai, E., Procaccia, S., and Seger, R. (2011). The MAPK cascades: signaling components, nuclear roles and mechanisms of nuclear translocation. *Biochimica et Biophysica Acta*, 1813(9), 1619–1633.doi.org/10.1016/j.bbamcr.2010.12.012.

Poncini, C. V., Soto, C. D. A., Batalla, E., Solana, M. E., and Gonzalez Cappa, S. M. (2008). *Trypanosoma cruzi* induces regulatory dendritic cells in vitro. *Infection and Immunity,* 76, 2633–2641. doi: 10.1128/IAI.01298-07.

Prudencio, M., Rodriguez, A., and Mota, M. M. (2006). The silent path to thousands of merozoites: the *Plasmodium* liver stage. *Nature Reviews Microbiology,* 4(11): 849-856. doi: 10.1038/nrmicro1529.

Ramphul, U. N., Garver, L. S., Molina-Cruz, A., Canepa, G. E., and Barillas-Mury C. (2015). *Plasmodium falciparum* evades mosquito immunity by disrupting JNK-mediated apoptosis of invaded midgut cells. *Proceedings of the National Academy of Sciences U S A*, 112(5), 1273–80. doi: 10.1073/pnas.1423586112.

Rodríguez-González, J., Wilkins-Rodríguez, A. A., Argueta-Donohué, J., and Gutiérrez-Kobeh, L. (2016). *Leishmania mexicana* promastigotes down regulate JNK and p-38 MAPK activation: Role in the inhibition of camptothecin-induced apoptosis of monocyte-derived dendritic cells. *Experimental Parasitology*, 163, 57–67. doi: 10.1016/j.exppara. 2015.12.005.

Ropert, C., Almeida, I. C., Closel, M., Travassos, L. R., Ferguson, M. A. J., Cohen, P., Gazzinelli, R. T. (2001). Requirement of mitogen-activated protein kinases and I B phosphorylation for induction of proinflammatory cytokines synthesis by macrophages indicates functional similarity of receptors triggered by glycosylphosphatidylinositol anchors from parasitic protozoa and bacterial lipopolysaccharide. *Journal of Immunology*, 166, 3423–3431. doi: 10.4049/jimmunol.166.5.3423.

Rubinfeld, H. and Seger, R. (2005). The ERK Cascade: A prototype of MAPK signaling. *Molecular Biotechnology*, 31(2), 151–174. doi: 10.1385/MB:31:2:151.

Ruhland, A. and Kima, P. E. (2009) Activation of PI3K/Akt signaling has a dominant negative effect on IL-12 production by macrophages infected with *Leishmania amazonensis* promastigotes. *Experimental Parasitology*, 122(1), 28–36. doi: 10.1016/j.exppara.2008.12.010.

Ruiz Díaz, P., Mucci, J., Meira, M. A., Bogliotti, Y., Musikant, D., Leguizamón, M. S., and Campetella, O. (2015). *Trypanosoma cruzi* trans-sialidase prevents elicitation of th1 cell response via interleukin 10 and downregulates th1 effector cells. *Infection and Immunity*, 83, 2099–2108. doi: 10.1128/IAI.00031-15.

Sarkar, A., Aga, E., Bussmeyer, U., Bhattacharyya, A., Möller, S., Hellberg, L., Behnen, M., Solbach, W. and Laskay, T. (2013). Infection of neutrophil granulocytes with *Leishmania major* activates ERK 1/2 and modulates multiple apoptotic pathways to inhibit apoptosis. *Medical Microbiology and Immunology*, 202(1), 25–35. doi:10.1007/s00430-012-0246-1.

Sassmann-Schweda, A., Singh, P., Tang, C., Wietelmann, A., Wettschureck, N. and Offermanns, S. (2016). Increased apoptosis and browning of TAK1-deficient adipocytes protect against obesity. *JCI Insight*, 1(7), e81175. doi: 10.1172/jci.insight.81175.

Schöneck, R., Plumas-Marty, B., Taibi, A., Billaut-Mulot, O., Loyens, M., Gras-Masse, H., Capron, A., and Ouaissi, A. (1994). Trypanosoma cruzi cDNA encodes a tandemly repeated domain structure characteristic of small stress proteins and glutathione S-transferases. *Biology of the Cell*, 80, 1–10. doi: 10.1016/0248-4900(94)90011-6.

Shweash, M., McGachy, A., Schroeder, J., Neamatallah, T., Bryant, C. E., Millington O., Mottram, J. C., Alexander, J. and Plevin, R. (2011). *Leishmania mexicana* promastigotes inhibit macrophage IL-12 production via TLR-4 dependent COX-2, iNOS and arginase-1 expression. *Molecular Immunology*, 48(15–16), 1800–1808. doi: 10.1016/j.molimm.2011.05.013.

Sui, X., Kong, N., Ye, L., Han, W., Zhou, J., Zhang, Q., He, C. and Pan, H. (2014). p38 and JNK MAPK pathways control the balance of apoptosis and autophagy in response to chemotherapeutic agents. *Cancer Letters*, 344(2), 174–179. doi.org/10.1016/j.canlet.2013.11.019.

Tafolla, E., Wang, S., Wong, B., Leong, J. and Kapila, Y. L. (2005). JNK1 and JNK2 oppositely regulate p53 in signaling linked to apoptosis triggered by an altered fibronectin matrix: JNK links FAK and p53. *The Journal of Biological Chemistry*, 280(20), 19992–19999. doi: 10.1074/jbc.M500331200.

Tarleton, R. L. (2007). Immune system recognition of *Trypanosoma cruzi*. *Current Opinion in Immunology*, 19, 430–434. doi: 10.1016/j.coi.2007.06.003.

Tena, G., Asai, T., Chiu, W. L. and Sheen, J. (2001). Plant mitogen-activated protein kinase signaling cascades. *Current Opinion in Plant Biology*, 4(5), 392–400. doi: 10.1016/S1369-5266(00)00191-6.

Terrazas, C. A., Huitron, E., Vazquez, A., Juarez, I., Camacho, G. M., Calleja, E. A., and Rodriguez-Sosa, M. (2011). MIF synergizes with *Trypanosoma cruzi* antigens to promote efficient dendritic cell maturation and IL-12 production via p38 MAPK. *International Journal of Biological Sciences*, 7, 1298–1310. doi: 10.7150/ijbs.7.1298.

Tibbles, L. A. and Woodgett, J. R. (1999). The stress-activated protein kinase pathways. *Cellular and Molecular Life Sciences (CMLS)*, 55(10), 1230–1254. doi.org/10.1007/s000180050369.

Valere, A., Garnotel, R., Villena, I., Guenounou, M., Pinon, J. M., and Aubert D (2003). Activation of the cellular mitogen-activated protein kinase pathways ERK, p38 and JNK during *Toxoplasma gondii* invasion. *Parasite* 10: 59–64. doi: 10.1051/parasite/2003101p59.

Vázquez-López, R., Argueta-Donohué, J., Wilkins-Rodríguez, A., Escalona-Montaño, A., and Gutiérrez-Kobeh, L. (2015). *Leishmania mexicana* amastigotes inhibit p38 and JNK and activate PI3K/AKT: role in the inhibition of apoptosis of dendritic cells. *Parasite Immunology*, 37(11), 579–589. doi.org/10.1111/pim.12275.

Wang, T., Zhou, J., Gan, X., Wang, H., Ding, X., Chen, L., Wang, Y., DU, J., Shen, J., and Yu, L. (2014). *Toxoplasma gondii* induces apoptosis of neural stem cells via endoplasmic reticulum stress pathway. *Parasitology*, 141(7), 988–995. doi: 10.1017/S0031182014000183.

Waskiewicz, A. J. and Cooper, J. A. (1995). Mitogen and stress response pathways: MAP kinase cascades and phosphatase regulation in mammals and yeast. *Current Opinion in Cell Biology*, 7(6), 798–805. doi.org/10.1016/0955-0674(95)80063-8.

World Health Organization. Chagas disease (American trypanosomiasis). *Fact sheet* No. 2015:340.

World Health Organization. *World malaria report* 2016. Geneva, Switzerland: 2017.

Zhou, J., Gan, X., Wang, Y., Zhang, X., Ding, X., Chen, L., Du, J., Luo, Q., Wang, T., Shen, J., and Yu, L. (2015). *Toxoplasma gondii* prevalent in China induce weaker apoptosis of neural stem cells C17.2 *via* endoplasmic reticulum stress (ERS) signaling pathways. *Parasites and Vectors*, 8, 73. doi: 10.1186/s13071-015-0670-3.

Chapter 2

ROLES OF MAPKs IN CIRCADIAN CLOCK REGULATION IN VERTEBRATES

Yoshimi Okamoto-Uchida[1,†], PhD, Junko Izawa[2,†], PhD and Jun Hirayama[2,*], PhD

[1]Division of Medicinal Safety Science,
National Institute of Health Sciences, Tokyo, Japan
[2]Department of Clinical Engineering, Faculty of Health Sciences,
Komatsu University, Ishikawa, Japan

ABSTRACT

Circadian clocks are intrinsic, time-tracking systems that endow organisms with a survival advantage. At the molecular level, they can be divided into three conceptual components. The first is the pacemaker, dedicated to generating and sustaining circadian rhythms in physiology by receiving and integrating signals from external time cues. The second component is the input, which refers to the pathway through which these cues are perceived and act upon the circadian pacemaker. The third element relates to how the clock affects physiology, which is achieved through the

---

[†] Equally contributed.
[*] Corresponding Author's E-mail: hirayamajun2003jp@yahoo.co.jp.

output pathways. There are three major MAPKs: c-JUN N-terminal kinase (JNK), p38, and extracellular signal-regulated kinase (ERK). Varieties of studies have revealed the critical roles of these MAPKs in regulating the pacemaker, input, and output pathways of circadian clocks. We describe selected regulatory aspects of circadian clocks in vertebrates, providing an intriguing link between the MAPKs and circadian clocks.

Keywords: ERK, p38, JNK, circadian clock, cellular clock

INTRODUCTION

Circadian clocks are endogenous oscillators that drive the daily rhythms of organisms ranging from bacteria to humans [1, 2]. These clocks regulate various biochemical, physiological, and behavioral processes with a periodicity of approximately 24 h. Under natural conditions, circadian clocks are entrained to this 24 h day by environmental time cues, such as light, to maximize an organism's physiological efficiency [3, 4]. Thus, the disruption of circadian clocks can have a profound effect on animal health and is linked to a variety of diseases, such as sleep disorders and metabolic syndrome [5, 6].

At the molecular level, circadian clocks can be divided into three conceptual components [7, 8]. The first is the pacemaker, dedicated to generating and sustaining rhythms by receiving and integrating signals from these cues. The second component is the input, which refers to the pathway through which external time cues are perceived and act upon the central pacemaker. The third element relates to how the clock affects physiology, which is achieved through the output pathways [9].

The pacemaker of circadian clocks is the cell-autonomous and self-sustained transcriptional machinery called the cellular clock, which is present in every single cell of living organisms [2, 10] (Figure 1). The appropriate synchronization of cellular clocks in tissues and organs is required for the generation of circadian rhythms in a variety of physiological processes, such as sleep and metabolism [11]. In most organisms, the molecular mechanisms underlying the establishment and maintenance of cellular clocks comprise interconnected transcription–translation feedback loops in which some clock factors repress their own transcription once they have attained critical levels [2, 7].

Phosphorylation has essential roles in appropriately regulating cellular clocks by modifying various biological processes at the molecular level, including gene expression, chromatin remodeling, and protein stabilization [12, 13]. In addition, a variety of studies have revealed that phosphorylation-mediated signaling pathways are involved in the regulation of the input and output processes of circadian clocks [14-16]. In particular, MAPKs have been shown to be important regulators of the above-mentioned phosphorylation-dependent controls. Here we describe several essential roles of MAPKs and their interaction with other kinases and phosphatases in the establishment of circadian clocks.

CELLULAR CLOCK REGULATION IN VERTEBRATES

In vertebrate cellular clocks, CLOCK or NPAS2 heterodimerizes with BMAL to form a transcriptionally active complex that facilitates expression of the Period (*Per*) and Cryptochrome (*Cry*) genes [2, 10, 17] (Figure 2). Once the PER and CRY proteins have been translated, they repress CLOCK (NPAS2): BMAL-mediated transcription, setting up the rhythmic gene expression that drives the circadian clock. The CLOCK (NPAS2):BMAL complex also stimulates expression of the clock-controlled genes (Ccgs) to regulate various elements of physiology. This accounts in part for the presence of circadian rhythms in a variety of physiological processes. A group of eight proteins comprises the basic sprockets of the molecular wheel that controls the mammalian circadian clock: PER1, PER2, PER3, CRY1, CRY2, CLOCK, NPAS2, BMAL2, and BMAL1 [2, 18]. Studies of these genes using mutant mice have revealed the distinct roles of clock proteins in regulating circadian clocks as well as direct links between the circadian clock and non-circadian aspects of animal physiology [6, 19, 20].

ROLE OF ERK/MAPK SIGNALING PATHWAY IN LIGHT-DEPENDENT REGULATION OF RETINAL CELLULAR CLOCKS

To guarantee that an organism's behavior remains tied to the rhythms of its environment, the circadian clocks must respond to environmental stimuli to be reset [3, 21]. The main cue for animals is light, which is provided by the day–night cycle. The eye is the principal mediator of light input to the central nervous system in mammals [22, 23]. The light signal to the circadian clock is integrated by a specific subset of cells, the retinal ganglion cells, localized in the ganglion cell layer of the retina.

Dopamine is the major catecholamine in the vertebrate retina and plays a central role in neural adaptation to light [24, 25]. Indeed, light stimulates the synthesis, turnover, and release of retinal dopamine, which makes dopamine an important mediator of light signaling to the retinal cellular clocks. Among members of the dopamine receptor family, the dopamine D2 receptor (D2R) has been shown to control light-induced reset of the circadian clock [26, 27]. At the molecular level, it has been reported that signaling mediated by the D2R enhances the transcriptional capacity of the CLOCK:BMAL complex. This effect involves the ERK/MAPK transduction cascade and is associated with a D2R-induced increase in the phosphorylation of the transcriptional coactivator cAMP-responsive element-binding protein (CREB) binding protein and its recruitment to the CLOCK:BMAL complex. Importantly, this activation of CLOCK:BMAL1-dependent transcription is responsible for the induction of the *Per1* gene by light in the retina, which is in turn responsible for the reset of the retinal cellular clock [26]. These findings provide evidence for the physiological links among the ERK/MAPK signaling pathway, dopamine, and the light input pathway of circadian clocks.

ROLES OF MAPKs IN LIGHT INPUT PATHWAYS IN CENTRAL CLOCKS IN MAMMALS

Photic information received in the retina is then conveyed through the retinohypothalamic tract to the suprachiasmatic nucleus (SCN), where the central clock is located [3, 28] (Figure 1). The involvement of the ERK/MAPK pathway in the light input system of the circadian clock in the SCN has been well established. Mice exposed to light only during their subjective night have shown a rapid and significant upregulation of the active phosphorylated form of ERK in the SCN [29]. Studies using transgenic and knockout mice of Raf kinase inhibitor protein, a suppressor of ERK activation, have clearly demonstrated that ERK activation coupled with photic input to regulate the SCN clock [30]. The light-activated ERK/MAPK pathway induces the phosphorylation of cAMP-responsive element (CRE)-binding protein (CREB), which includes expression of clock components *Per1* and *Per2 in the SCN* [31, 32]. This event is thought to contribute to phase shifting of the circadian clock at the animal level [33, 34]. However, whether light-dependent induction of *Per1* and *Per2* in the SCN is involved with the phase shifting has not been fully elucidated using adequate genetically modified mice. Mouse *Per1* and *Per2* genes are induced by the CLOCK (NPAS2):BMAL complex and by light. In particular, the CLOCK (NPAS2):BMAL-dependent regulation of *Per1* and *Per2* is essential for establishment of the circadian clock's rhythmicity [35, 36]. Thus, genetic inhibition of both mouse *Per1* and *Per2* genes disrupts the formation of the circadian clock itself and thus prevents the analysis of light-dependent changes in circadian clock phase. This problem was overcome in the recent study using a zebrafish model described below.

ZEBRAFISH IS A USEFUL MODEL OF LIGHT-DEPENDENT REGULATION OF CIRCADIAN CLOCKS

The zebrafish is a particularly useful animal model for studying the light-dependent regulation of the circadian clock [37, 38]. In zebrafish, there are four transcriptional repressor type *Cry* genes (*zCry1a*, *zCry1b*, *zCry2a*, and *zCry2b*) and four *Per* genes (*zPer1a*, *zPer1b*, *zPer2*, and *zPer3*) [39-42] (Figure 3). *zCry2a* and *zCry2b* are induced by both the CLOCK (NPAS2):BMAL complex and by light; *zCry1b*, *zPer1a*, *zPer1b*, and *zPer3* are induced by the CLOCK (NPAS2):BMAL complex but not by light; and *zCry1a* and *Per2* are induced by light but not by the CLOCK (NPAS2):BMAL complex [43-48]. These distinct dependencies of *zPer* and *zCry* gene expression allowed analyses of the role of light-induced cellular clock regulators including zPER2, zCRY1a, and zCRY2a in the light-dependent regulation of circadian clocks [48].

The formation of behavioral rhythms in zebrafish is a useful experimental system to analyze light-dependent regulation of the circadian clock at the animal level [49]. During zebrafish development, organogenesis is completed within *two days post fertilization (dpf)*. Zebrafish larvae hatch within four dpf and then start to display locomotor behavior. Zebrafish cellular clocks are autonomously set in motion during development within 1–4 dpf, but are out of phase with each other in tissues and organs [46, 50]. Light synchronizes the phases of the cellular clocks to establish behavioral rhythms. A recent study generated *zCry1a$^{-/-}$ zPer2$^{-/-}$ zCry2a$^{-/-}$ triple KO (TKO) zebrafish and found that these zebrafish have a defect in light-dependent behavior rhythm formation during development [48]. Thus, zPER2, zCRY1a, and zCRY2a help to synchronize cellular clocks in a light-dependent manner, thus contributing to behavioral rhythm formation in zebrafish (Figure 3).*

MAPKS ARE AN ESSENTIAL REGULATOR OF LIGHT RESPONSES OF THE CELLULAR CLOCK IN ZEBRAFISH

Cultured zebrafish cells provide a valuable tool for studying the light-dependent regulation of cellular clocks [37, 43, 51]. In these cell lines, light induces the expression of clock genes including *zPer2, zCry1a, and zCry2a, and* the oscillations of clock-controlled gene expression can be entrained to a new light–dark cycle [52-54]. These facts show that cultured zebrafish cells have the clock components required for a light-induced resetting of their cellular clocks.

Studies using cultured zebrafish cells have identified important roles of three major MAPKs, JNK, p38, and ERK, in transducing light signals to the cellular clock machinery. Light has been reported to activate MAPK signaling cascades in zebrafish cells [55-57] (Figure 3). By utilizing a pharmacological approach, it was established that the light-induced *zPer2* transactivation requires the ERK/MAPK signaling pathway [40, 54]. It has also been proposed that light-induced ERK activation triggers *zCry1a* transcription, whereas light-induced p38 activation suppresses expression of *zCry1a* and *zPer2*, highlighting a MAPK-mediated cross-regulatory mechanism of the expression of circadian clock genes [40, 55]. In contrast to these studies, it has been recently reported that the light-activated p38 pathway facilitates expression of *zCry1a* and *zPer2* and that the ERK/MAPK signaling pathway is not involved in light-induced expression of *zCry1a* and *zPer2* [57]. It should be noted that the reason the contradictory results exist is unknown. In addition, it was reported that the light-activated JNK signaling pathway induces expression of *zCry1a* and *zPer2* [57].

LIGHT-INDUCED CHANGES IN CELLULAR REDOX STATE ACTIVATES MAPK SIGNALING CASCADES IN ZEBRAFISH

An important issue concerns the identity of the molecules responsible for light-dependent activation of MAPKs in the zebrafish system. In a variety of organisms, light induces reactive oxygen species (ROS) production that alters the redox state within cells [58, 59]. This light-induced change in redox state has been reported to stimulate intracellular MAPK signaling pathways in zebrafish cells and transduce photic signals to circadian transcription [56, 60]. In addition, several groups have identified flavin-containing oxidases as the phototransducing molecules responsible for light-dependent ROS production [57, 59, 61]. They are versatile flavoenzymes that catalyze molecular oxidation in numerous metabolic pathways, generating ROS as a byproduct. The absorbance of light in the near violet–blue region by these enzymes activates them and induces the photoreduction of the flavin adenine dinucleotide (FAD) moiety, thus leading to ROS production in a light-dependent manner [59, 62]. Notably, this cellular ROS production has been linked to the toxic effects of photo-oxidative stress in living organisms [58, 63]. In addition, MAPKs function in cellular responses to varieties of stresses, including radiation, UV, and oxidative stress [64]. These facts are consistent with the proposal that cellular responses to photo-oxidative stress are the evolutionary origin of circadian clocks. It is tempting to speculate that, in zebrafish, the light response of the cellular clock reflects a long-standing cellular response to photo-oxidative stress.

PHOSPHORYLATION IS AN IMPORTANT REGULATORY MECHANISM OF CELLULAR CLOCK PERIODICITY

It is currently well-accepted that the orchestrated temporal program of clock protein phosphorylation contributes to the proper regulation of vertebrate circadian clocks [15]. In particular, MAPKs are important for the circadian oscillation of phosphorylated clock proteins and are required for the maintenance of cellular clock regulation [65]. Additionally, a number of other kinases, such as casein kinase (CK) 1, CK2, and glycogen synthase kinase (GSK), have also been implicated as important regulators of cellular clocks and interact with MAPKs to establish circadian periodicity of these clocks [15, 66-69]. Further, participation of phosphatases in the circadian time–dependent phosphorylation of cellular clock regulators has also been well-recognized [16, 70, 71]. It is therefore highly probable that multiple kinases and phosphatases specifically regulate the functions of cellular clock regulators by generating daily changes in phosphorylation and dephosphorylation on each of them.

Besides phosphorylation, other posttranslational modifications, including acetylation, methylation, ADP-ribosylation and SUMOylation, are important modifications of varieties of cellular clock regulators, regulating their transcriptional activity, subcellular localization, and protein stability [15, 20, 72-74]. Thus, these modifications also play a role in controlling the oscillatory mechanism itself and/or in modulating the cellular clock light signaling pathway. It should be noted that they have also been proposed to interact with phosphrylation induced by MAPKs for the regulation of cellular clocks.

ROLES OF JNK IN REGULATION OF THE PACEMAKER OF CIRCADIAN CLOCKS

The functions of various cellular clock regulators, including CLOCK, BMAL1, PER1, PER2, CRY1, and CRY2, are regulated via phosphorylation by various enzymes, including CK1ϵ, CK1δ, GSK3β, CK2, and MAPKs [15, 16]. Genetic studies have revealed the importance of this phosphorylation in mammalian circadian clock regulation [72]. For example, in the Syrian hamster, the *tau* mutation causing a short-period phenotype affects the gene encoding CK1ϵ. CK1ϵ was subsequently demonstrated to phosphorylate

PER2, and the short-period phenotype of *tau* hamsters was directly linked to their lower rate of CK1ϵ-dependent PER2 phosphorylation [75]. Intriguingly, a defect in CK1ϵ-mediated PER2 phosphorylation has also been implicated in human sleep disorders [76]. Indeed, familial advanced sleep phase syndrome (FASPS) has been associated with a missense mutation in the *Per2* gene and the produced mutant protein is less effectively phosphorylated by CKIε *in vitro*. Furthermore, a polymorphism in a region of the human *per3* gene, the presumed CKIε binding domain, has been suggested to be associated with a delayed sleep phase syndrome [77]. At the molecular level, CK1ϵ-mediated PER2 phosphorylation has been shown to decrease the stability of PER2 protein by promoting its ubiquitination [78, 79]. Notably, changes in PER2 stability have been linked to changes in the period length of cellular clocks.

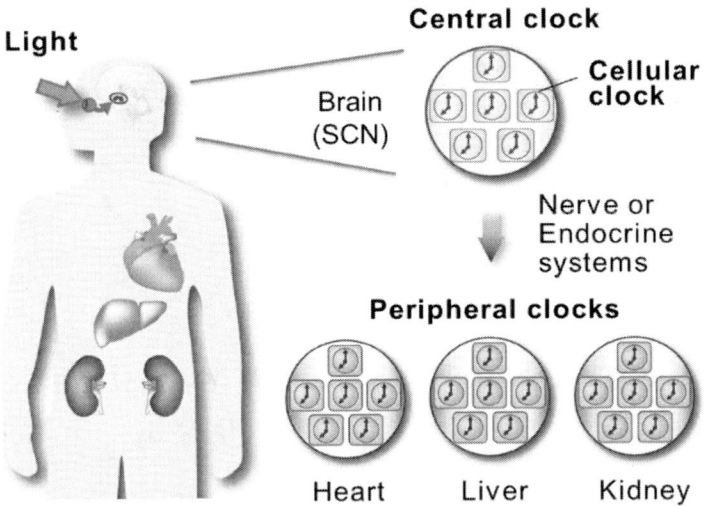

Figure 1. Circadian clock system in mammals. The circadian clocks reset themselves in response to light, which is provided in day-night cycles. Both vertebrates and invertebrates have circadian oscillators, cellular clocks, scattered throughout their bodies. In mammals, the circadian system is composed of both central and peripheral clocks. The mammalian central clock is located in the suprachiasmatic nucleus (SCN) within the anterior hypothalamus of the brain and integrates photic cues from the retina. This central clock acts as a coordinator and provides time signals via both neural and humoral routes that entrain independent peripheral clocks.

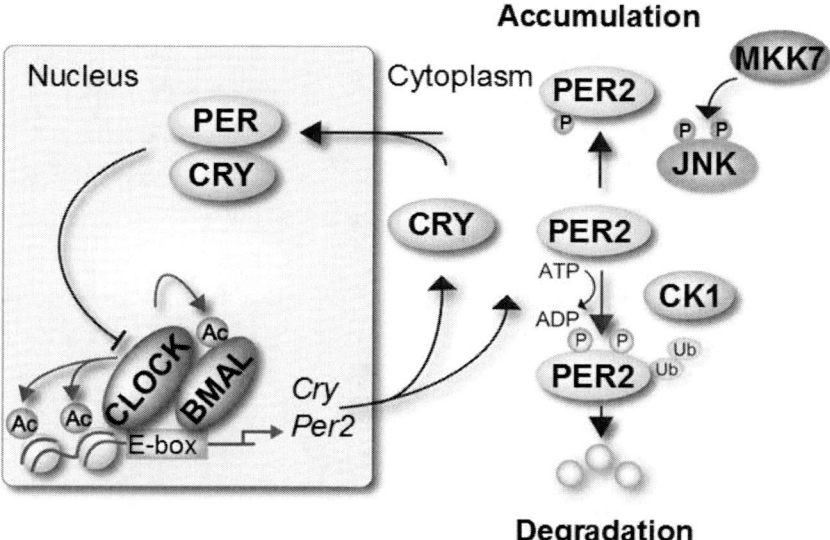

Figure 2. Post-translational modification of cellular clock regulators. Post-translational modifications, such as phosphorylation and acetylation, are important modulators of cellular clock components and regulate their transcriptional activity, subcellular localization and stability. In mammals, the core circadian regulator CLOCK has intrinsic histone acetyltransferase (HAT) activity that it uses to acetylate its heterodimeric partner BMAL (99). This CLOCK-mediated acetylation increases the interaction of the CLOCK:BMAL complex with CRY (60). PER2 is an essential circadian component, whose functions are regulated by a variety of kinases including CK1 and JNK. MKK7-mediated JNK activation induces phosphorylation of PER2 and increases its protein half-life by inhibiting its ubiquitination. Notably, the MKK7/JNK-dependent stabilization of PER2 competes with the CK1-induced PER2 degradation. Abbreviations; P: Phosphorylation, Ac: Acetylation, Ub: Ubiquitination.

JNK, a member of the MAPK family, is activated by many types of external stresses, including changes in osmolarity, heat shock, and UV irradiation; this activity is regulated via the phosphorylation of particular tyrosine and threonine residues located in the kinase domain [80]. JNK phosphorylation is catalyzed by two dual-specificity kinases, MKK4 and MKK7, that act in a synergistic manner [81]. Although most often activated in response to stress, phosphorylated JNK has been detected in unstressed cultured cells and in isolated mouse tissues, such as the brain [82]. These facts indicate the importance of JNK signaling in physiological processes other than cellular stress responses. In fact, it has been shown that the JNK signaling pathway is an essential regulator of the periodicity of cellular clocks; in other words, the circadian pacemaker [83, 84]. It has been reported that the MKK7–JNK signaling pathway induces phosphorylation of PER2 and stabilizes PER2 by inhibiting its ubiquitination (Figure 2). In fact, the MKK7–JNK signaling pathway has an effect opposite to that of CK1ϵ-induced PER2 destabilization, providing a balancing influence on clock protein functions that helps to maintain the normal periodicity of the circadian clock machinery. In addition, the JNK signaling pathway has been reported to phosphorylate BMAL1 and CLOCK in mice [83, 85]. In particular, Yoshitane et al. have clearly shown that JNK-mediated phosphorylation of BMAL1 is an important regulatory mechanism of light-dependent regulation of the mammalian circadian clock *in vivo* [85].

ROLES OF ERK IN FUNCTIONAL CONTROLS OF CELLULAR CLOCK REGULATORS

It has been reported that ERK directly interacts with and phosphorylates mouse CRY1 and CRY2 (86). Ser247 of mouse CRY1 and the corresponding residue Ser265 of mouse CRY2, which are highly conserved among vertebrate CRYs, have been shown to constitute the phosphorylation target site of ERK *in vitro*. Phosphorylation of these residues has been suggested to reduce CRY's repressor activity. MAPK also phosphorylates the well-conserved Ser557 in the mouse CRY2 C-terminal domain. The biological importance of this phosphorylation of mammalian CRYs remains to be determined.

Two kinases, ERK and CKI, have been shown to phosphorylate BMAL1 protein *in vitro* [87, 88]. ERK phosphorylates three mouse BMAL1 residues, Ser-527, Thr-534, and Ser-599, residues, which are highly conserved among vertebrate BMAL1s. Vertebrate BMAL1s also have a well-conserved CKI phosphorylation motif in their C-terminal domain [15]. Interestingly, this putative CKI phosphorylation motif overlaps in part with one of ERK, indicative that both kinases could compete to regulate BMAL1 function.

Consistent with this idea, an *in vitro* transcriptional assay showed that ERK phosphorylation of mouse BMAL1 downregulates CLOCK:BMAL1 transcription, whereas CKI phosphorylation has been reported to facilitate it [87, 88].

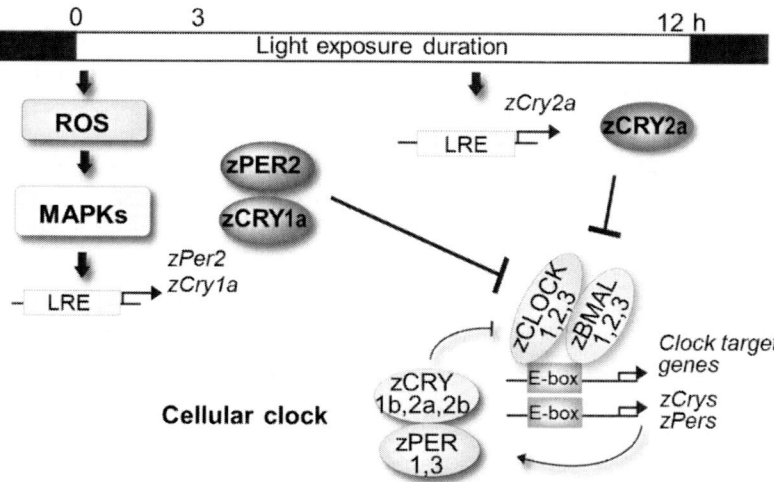

Figure 3. Molecular mechanisms of light-dependent regulation of zebrafish cellular clocks. Zebrafish possess an intrinsic circadian oscillator comprising components similar to those of mammals. Zebrafish CLOCK1, CLOCK2 (NPAS2), CLOCK3, BMAL1, BMAL2, and BMAL3 are transcriptional activators and zebrafish PER1, PER2, PER3, CRY1a, CRY1b, CRY2a, and CRY2b are transcriptional repressors. *zPer2* and *zCry1a* expression occurs in a light stimulation-dependent manner, whereas that of *zPer1*, *zPer3*, *zCry1b*, *zCry2a*, and *zCry2b* is under the control of the heterodimeric zCLOCK (zNPAS2):zBMAL transcription factors, which bind to E-box elements in the promoters of target genes. In addition, *zCry2a* is induced by both the zCLOCK (zNPAS2):zBMAL complex and by light. Light induced expression of *zCry2a* reaches high levels at 12-h after light onset, whereas maximum expression levels of *zCry1a* and *zPer2* are observed at 3-h after light onset (48). Light drives the production of intracellular ROS. Light-induced ROS can perform signaling roles by stimulating MAPKs, which lead to transcriptional activation of the *zCry1a* and *zPer2* genes. On the other hand, the signaling cascades for the light induction of *zCry2a* has not been identified. Light-induced zPER2, zCRY1a and zCRY2a inhibit transcription in a zCLOCK (zNPAS2):zBMAL-dependent manner, thereby contributing to the light-dependent regulation of cellular clocks, such as their synchronization.
Abbreviations; ROS: **Reactive** oxygen species, LRE: Light responsive element.

ROLES OF ERK IN SYNCHRONIZATION OF PERIPHERAL CELLULAR CLOCKS

In mammals, light signals are received by the retina and integrated to the SCN cellular clock. The SCN cellular clock then transmits information regarding the light to peripheral cellular clocks via humoral signals, which resets them to induce their synchronization [3, 4]. Interestingly, recent studies have reported that factor(s) other than cellular clocks in the SCN can synchronize peripheral cellular clocks in a light-dependent manner [89]. Furthermore, peripheral clocks can also respond directly to SCN-independent signals such as feeding and temperature changes to synchronize to each other [8]. A variety of studies have reported that the ERK/MAPK signaling pathway is involved in resetting and synchronizing the peripheral clocks. For example, there is considerable evidence that the ERK/MAPK pathway is involved in insulin-mediated circadian phase shift in insulin-sensitive peripheral tissues. Firstly, Akashi et al. have provided solid evidence that the ERK/MAPK signaling pathway resets cellular clocks by using mouse NIH3T3 cultured cells, a well-established *in vitro* system for analysis of peripheral cellular clocks [90]. Secondly, insulin has been shown to activate the MAPK signaling pathway to induce expression of the *Per2* clock gene in H4IIE rat hepatoma cells [91]. It is well established that insulin activates the insulin receptor, which leads to cellular glucose uptake and activation of the ERK/MAPK signaling pathway. Thirdly, based on a pharmacological approach, this insulin-induced activation of the ERK/MAPK pathway has been reported to be required for insulin-mediated reset of cellular clocks in white adipose tissue [92]. Feeding facilitates insulin levels in the body and resets the circadian clock in peripheral tissues and organs [8]. Thus, it is conceivable that feeding causes circadian phase shift of peripheral cellular clocks by the induction of insulin, which activates the ERK/MAPK cascade to increase the expression of the clock gene *Per2*.

ROLES OF P38 IN THE REGULATION OF CELLULAR CLOCK PERIODICITY

It has been reported that the activity of p38 kinase does not show a circadian oscillation in the chicken pineal gland [93]. Cellular clocks in the pineal gland establish a circadian rhythm by producing melatonin, a sleep-inducing hormone. Notably, pharmacological inhibition of p38 kinase activity was shown to lengthen the free-running period of the melatonin secretion rhythm in cultured chick pineal cells [93]. Consistent with this report, the periodicity of cellular clocks is lengthened in mammalian cultured cells by treatment with a p38 inhibitor, SB203580 [94]. Thus, p38 kinase is thought to be involved in the maintenance of proper periodicity in cellular clocks.

POSSIBLE ROLES OF MAPKS IN THE REGULATION OF THE CIRCADIAN CLOCK OUTPUT PATHWAY

It should be noted that the activity of MAPKs including ERK and JNK shows circadian oscillation in mouse peripheral and SCN cells [95-97]. The levels of activated MAPKs are significantly reduced in cellular clock-deficient cells, indicating that the circadian clock plays an important role in activation of these pathways. MAPKs have been shown to be involved in a variety of physiological processes such as hormonal synthesis and release, body temperature control, and further aspects of metabolism [81, 98]. Thus, it is conceivable that circadian clocks generate daily rhythms of these physiologies via regulation of activities of MAPKs.

ACKNOWLEDGMENTS

This work was supported in part by the Japan Society for the Promotion of Science Grant-in-Aid for Scientific Research [16K08521 and 18KT0068

(J. H.)]. This work was also supported by grants from the Magnetic Health Science Foundation (J. H.), the *Ichiro Kanehara Foundation (J. H.) and Komatsu University* (J. I. and J. H.).

REFERENCES

[1] Fisher, S. P., Foster, R. G. and Peirson, S. N. (2013) The circadian control of sleep. *Handb Exp Pharmacol*, 157-183.
[2] Takahashi, J. S. (2017) Transcriptional architecture of the mammalian circadian clock. *Nat Rev Genet*, **18**, 164-179.
[3] King, D. P. and Takahashi, J. S. (2000) Molecular genetics of circadian rhythms in mammals. *Annu Rev Neurosci*, **23**, 713-742.
[4] Harmer, S. L., Panda, S. and Kay, S. A. (2001) Molecular bases of circadian rhythms. *Annu Rev Cell Dev Biol*, **17**, 215-253.
[5] Kondratov, R. V., Gorbacheva, V. Y. and Antoch, M. . (2007) The role of mammalian circadian proteins in normal physiology and genotoxic stress responses. *Curr Top Dev Biol*, **78**, 173-216.
[6] Sahar, S. and Sassone-Corsi, P. (2009) Metabolism and cancer: the circadian clock connection. *Nat Rev Cancer*, **9**, 886-896.
[7] Dunlap, J. C. (1999) Molecular bases for circadian clocks. *Cell*, **96**, 271-290.
[8] Schibler, U. and Sassone-Corsi, P. (2002) A web of circadian pacemakers. *Cell*, **111**, 919-922.
[9] Cermakian, N. and Sassone-Corsi, P. (2000) Multilevel regulation of the circadian clock. *Nature reviews. Molecular cell biology*, **1**, 59-67.
[10] Panda, S., Hogenesch, J. B. and Kay, S. A. (2002) Circadian rhythms from flies to human. *Nature*, **417**, 329-335.
[11] Dibner, C., Schibler, U. and Albrecht, U. (2010) The mammalian circadian timing system: organization and coordination of central and peripheral clocks. *Annu Rev Physiol*, **72**, 517-549.
[12] Whitmarsh, A. J. (2007) Regulation of gene transcription by mitogen-activated protein kinase signaling pathways. *Biochimica et biophysica acta*, **1773**, 1285-1298.

[13] Bigeard, J., Rayapuram, N., Pflieger, D. and Hirt, H. (2014) Phosphorylation-dependent regulation of plant chromatin and chromatin-associated proteins. *Proteomics*, **14**, 2127-2140.

[14] Widmann, C., Gibson, S., Jarpe, M. B. and Johnson, G. L. (1999) Mitogen-activated protein kinase: conservation of a three-kinase module from yeast to human. *Physiological reviews*, **79**, 143-180.

[15] Hirayama, J. and Sassone-Corsi, P. (2005) Structural and functional features of transcription factors controlling the circadian clock. *Curr Opin Genet Dev*, **15**, 548-556.

[16] Reischl, S. and Kramer, A. (2011) Kinases and phosphatases in the mammalian circadian clock. *FEBS letters*, **585**, 1393-1399.

[17] Sancar, A. (2003) Structure and function of DNA photolyase and cryptochrome blue-light photoreceptors. *Chem Rev*, **103**, 2203-2237.

[18] Matsumura, R. and Akashi, M. (2018) Role of the clock gene Period3 in the human cell-autonomous circadian clock. *Genes to cells : devoted to molecular & cellular mechanisms*.

[19] Kondratov, R. V., Kondratova, A. A., Gorbacheva, V. Y., Vykhovanets, O. V. and Antoch, M. P. (2006) Early aging and age-related pathologies in mice deficient in BMAL1, the core componentof the circadian clock. *Genes Dev*, **20**, 1868-1873.

[20] Uchida, Y., Hirayama, J. and Nishina, H. (2010) A common origin: signaling similarities in the regulation of the circadian clock and DNA damage responses. *Biol Pharm Bull*, **33**, 535-544.

[21] Sancar, A. (2000) Cryptochrome: the second photoactive pigment in the eye and its role in circadian photoreception. *Annu Rev Biochem*, **69**, 31-67.

[22] LeGates, T. A., Fernandez, D. C. and Hattar, S. (2014) Light as a central modulator of circadian rhythms, sleep and affect. *Nature reviews. Neuroscience*, **15**, 443-454.

[23] Besharse, J. C. and McMahon, D. G. (2016) The Retina and Other Light-sensitive Ocular Clocks. *Journal of biological rhythms*, **31**, 223-243.

[24] Korshunov, K. S., Blakemore, L. J. and Trombley, P. Q. (2017) Dopamine: A Modulator of Circadian Rhythms in the Central Nervous System. *Frontiers in cellular neuroscience*, **11**, 91.

[25] Solinas, M., Belujon, P., Fernagut, P. O., Jaber, M. and Thiriet, N. (2018) Dopamine and addiction: what have we learned from 40 years of research. *Journal of neural transmission* (Vienna, Austria : 1996).

[26] Yujnovsky, I., Hirayama, J., Doi, M., Borrelli, E. and Sassone-Corsi, P. (2006) Signaling mediated by the dopamine D2 receptor potentiates circadian regulation by CLOCK:BMAL1. *Proceedings of the National Academy of Sciences of the United States of America*, **103**, 6386-6391.

[27] Doi, M., Yujnovsky, I., Hirayama, J., Malerba, M., Tirotta, E., Sassone-Corsi, P. and Borrelli, E. (2006) Impaired light masking in dopamine D2 receptor-null mice. *Nat Neurosci*, **9**, 732-734.

[28] Okamura, H. (2004) Clock genes in cell clocks: roles, actions, and mysteries. *Journal of biological rhythms*, **19**, 388-399.

[29] Obrietan, K., Impey, S. and Storm, D. R. (1998) Light and circadian rhythmicity regulate MAP kinase activation in the suprachiasmatic nuclei. *Nat Neurosci*, **1**, 693-700.

[30] Antoun, G., Bouchard-Cannon, P. and Cheng, H. Y. (2012) Regulation of MAPK/ERK signaling and photic entrainment of the suprachiasmatic nucleus circadian clock by Raf kinase inhibitor protein. *J Neurosci*, **32**, 4867-4877.

[31] Zylka, M. J., Shearman, L. P., Weaver, D. R. and Reppert, S. M. (1998) Three period homologs in mammals: differential light responses in the suprachiasmatic circadian clock and oscillating transcripts outside of brain. *Neuron*, **20**, 1103-1110.

[32] Serchov, T. and Heumann, R. (2017) Ras Activity Tunes the Period and Modulates the Entrainment of the Suprachiasmatic Clock. *Frontiers in neurology*, **8**, 264.

[33] Shigeyoshi, Y., Taguchi, K., Yamamoto, S., Takekida, S., Yan, L., Tei, H., Moriya, T., Shibata, S., Loros, J. J., Dunlap, J. C. et al. (1997) Light-induced resetting of a mammalian circadian clock is associated with rapid induction of the mPer1 transcript. *Cell*, **91**, 1043-1053.

[34] Okamura, H., Miyake, S., Sumi, Y., Yamaguchi, S., Yasui, A., Muijtjens, M., Hoeijmakers, J. H. and van der Horst, G. T. (1999) Photic induction of mPer1 and mPer2 in cry-deficient mice lacking a biological clock. *Science* (New York, N.Y.), **286**, 2531-2534.

[35] Bae, K., Jin, X., Maywood, E. S., Hastings, M. H., Reppert, S. M. and Weaver, D. R. (2001) Differential functions of mPer1, mPer2, and mPer3 in the SCN circadian clock. *Neuron*, **30**, 525-536.

[36] Zheng, B., Albrecht, U., Kaasik, K., Sage, M., Lu, W., Vaishnav, S., Li, Q., Sun, Z. S., Eichele, G., Bradley, A. et al. (2001) Nonredundant roles of the mPer1 and mPer2 genes in the mammalian circadian clock. *Cell*, **105**, 683-694.

[37] Tamai, T. K., Carr, A. J. and Whitmore, D. (2005) Zebrafish circadian clocks: cells that see light. *Biochem Soc Trans*, **33**, 962-966.

[38] Idda, M. L., Bertolucci, C., Vallone, D., Gothilf, Y., Sanchez-Vazquez, F. J. and Foulkes, N. S. (2012) Circadian clocks: lessons from fish. *Prog Brain Res*, **199**, 41-57.

[39] Kobayashi, Y., Ishikawa, T., Hirayama, J., Daiyasu, H., Kanai, S., Toh, H., Fukuda, I., Tsujimura, T., Terada, N., Kamei, Y. et al. (2000) Molecular analysis of zebrafish photolyase/cryptochrome family: two types of cryptochromes present in zebrafish. *Genes to cells : devoted to molecular & cellular mechanisms*, **5**, 725-738.

[40] Cermakian, N., Pando, M. P., Thompson, C. L., Pinchak, A. B., Selby, C. P., Gutierrez, L., Wells, D. E., Cahill, G. M., Sancar, A. and Sassone-Corsi, P. (2002) Light induction of a vertebrate clock gene involves signaling through blue-light receptors and MAP kinases. *Curr Biol*, **12**, 844-848.

[41] Hirayama, J., Fukuda, I., Ishikawa, T., Kobayashi, Y. and Todo, T. (2003) New role of zCRY and zPER2 as regulators of sub-cellular distributions of zCLOCK and zBMAL proteins. *Nucleic Acids Res*, **31**, 935-943.

[42] Pando, M. P. and Sassone-Corsi, P. (2002) Unraveling the mechanisms of the vertebrate circadian clock: zebrafish may light the way. *BioEssays : news and reviews in molecular, cellular and developmental biology*, **24**, 419-426.

[43] Pando, M. P., Pinchak, A. B., Cermakian, N. and Sassone-Corsi, P. (2001) A cell-based system that recapitulates the dynamic light-dependent regulation of the vertebrate clock. *Proceedings of the National Academy of Sciences of the United States of America*, **98**, 10178-10183.

[44] Hirayama, J., Cardone, L., Doi, M. and Sassone-Corsi, P. (2005) Common pathways in circadian and cell cycle clocks: light-dependent activation of Fos/AP-1 in zebrafish controls CRY-1a and WEE-1. *Proceedings of the National Academy of Sciences of the United States of America*, **102**, 10194-10199.

[45] Tamai, T. K., Young, L. C. and Whitmore, D. (2007) Light signaling to the zebrafish circadian clock by Cryptochrome 1a. *Proceedings of the National Academy of Sciences of the United States of America*, **104**, 14712-14717.

[46] Dekens, M. P. and Whitmore, D. (2008) Autonomous onset of the circadian clock in the zebrafish embryo. *Embo J*.

[47] Li, Y., Li, G., Wang, H., Du, J. and Yan, J. (2013) Analysis of a gene regulatory cascade mediating circadian rhythm in zebrafish. *PLoS Comput Biol*, **9**, e1002940.

[48] Hirayama, J., Alifu, Y., Hamabe, R., Yamaguchi, S., Tomita, J., Maruyama, Y., Asaoka, Y., Nakahama, K. I., Tamaru, T., Takamatsu, K. et al. (2019) The clock components Period2, Cryptochrome1a, and Cryptochrome2a function in establishing light-dependent behavioral rhythms and/or total activity levels in zebrafish. *Scientific reports*, **9**, 196.

[49] Hurd, M. W. and Cahill, G. M. (2002) Entraining signals initiate behavioral circadian rhythmicity in larval zebrafish. *Journal of biological rhythms*, **17**, 307-314.

[50] Kaneko, M. and Cahill, G. M. (2005) Light-dependent development of circadian gene expression in transgenic zebrafish. *PLoS Biol*, **3**, e34.

[51] Hirayama, J., Kaneko, M., Cardone, L., Cahill, G. and Sassone-Corsi, P. (2005) Analysis of circadian rhythms in zebrafish. *Methods Enzymol*, **393**, 186-204.

[52] Whitmore, D., Foulkes, N. S. and Sassone-Corsi, P. (2000) Light acts directly on organs and cells in culture to set the vertebrate circadian clock. *Nature*, **404**, 87-91.

[53] Vallone, D., Gondi, S. B., Whitmore, D. and Foulkes, N. S. (2004) E-box function in a period gene repressed by light. *Proceedings of the National Academy of Sciences of the United States of America*, **101**, 4106-4111.

[54] Hirayama, J., Cho, S. and Sassone-Corsi, P. (2007) Circadian control by the reduction/oxidation pathway: catalase represses light-dependent clock gene expression in the zebrafish. *Proceedings of the National Academy of Sciences of the United States of America*, **104**, 15747-15752.

[55] Hirayama, J., Miyamura, N., Uchida, Y., Asaoka, Y., Honda, R., Sawanobori, K., Todo, T., Yamamoto, T., Sassone-Corsi, P. and Nishina, H. (2009) Common light signaling pathways controlling DNA repair and circadian clock entrainment in zebrafish. *Cell Cycle*, **8**, 2794-2801.

[56] Uchida, Y., Shimomura, T., Hirayama, J. and Nishina, H. (2011) Light, reactive oxygen species, and magnetic fields activate ERK/MAPK signaling pathway in cultured zebrafish cells. *Appl Magn Reson*, **42**, 69-77.

[57] Pagano, C., Siauciunaite, R., Idda, M. L., Ruggiero, G., Ceinos, R. M., Pagano, M., Frigato, E., Bertolucci, C., Foulkes, N. S. and Vallone, D. (2018) Evolution shapes the responsiveness of the D-box enhancer element to light and reactive oxygen species in vertebrates. *Scientific reports*, **8**, 13180.

[58] Neill, S. J., Desikan, R., Clarke, A., Hurst, R. D. and Hancock, J. T. (2002) Hydrogen peroxide and nitric oxide as signalling molecules in plants. *J Exp Bot*, **53**, 1237-1247.

[59] Hockberger, P. E., Skimina, T. A., Centonze, V. E., Lavin, C., Chu, S., Dadras, S., Reddy, J. K. and White, J. G. (1999) Activation of flavin-containing oxidases underlies light-induced production of H2O2 in mammalian cells. *Proceedings of the National Academy of Sciences of the United States of America*, **96**, 6255-6260.

[60] Hirayama, J., Sahar, S., Grimaldi, B., Tamaru, T., Takamatsu, K., Nakahata, Y. and Sassone-Corsi, P. (2007) CLOCK-mediated acetylation of BMAL1 controls circadian function. *Nature*, **450**, 1086-1090.

[61] Osaki, T., Uchida, Y., Hirayama, J. and Nishina, H. (2011) Diphenyleneiodonium chloride, an inhibitor of reduced nicotinamide adenine dinucleotide phosphate oxidase, suppresses light-dependent induction of clock and DNA repair genes in zebrafish. *Biol Pharm Bull*, **34**, 1343-1347.

[62] Hancock, J. T., Desikan, R. and Neill, S. J. (2001) Role of reactive oxygen species in cell signalling pathways. *Biochem Soc Trans*, **29**, 345-350.

[63] Neill, S., Desikan, R. and Hancock, J. (2002) Hydrogen peroxide signalling. *Curr Opin Plant Biol*, **5**, 388-395.

[64] Kyriakis, J. M. and Avruch, J. (2012) Mammalian MAPK signal transduction pathways activated by stress and inflammation: a 10-year update. *Physiological reviews*, **92**, 689-737.

[65] Goldsmith, C. S. and Bell-Pedersen, D. (2013) Diverse roles for MAPK signaling in circadian clocks. *Advances in genetics*, **84**, 1-39.

[66] Iitaka, C., Miyazaki, K., Akaike, T. and Ishida, N. (2005) A role for glycogen synthase kinase-3beta in the mammalian circadian clock. *The Journal of biological chemistry*, **280**, 29397-29402.

[67] Tamaru, T., Hirayama, J., Isojima, Y., Nagai, K., Norioka, S., Takamatsu, K. and Sassone-Corsi, P. (2009) CK2alpha phosphorylates BMAL1 to regulate the mammalian clock. *Nat Struct Mol Biol*, **16**, 446-448.

[68] Tsuchiya, Y., Akashi, M., Matsuda, M., Goto, K., Miyata, Y., Node, K. and Nishida, E. (2009) Involvement of the protein kinase CK2 in the regulation of mammalian circadian rhythms. *Science signaling*, **2**, ra26.

[69] Maier, B., Wendt, S., Vanselow, J. T., Wallach, T., Reischl, S., Oehmke, S., Schlosser, A. and Kramer, A. (2009) A large-scale functional RNAi screen reveals a role for CK2 in the mammalian circadian clock. *Genes Dev*, **23**, 708-718.

[70] Doi, M., Cho, S., Yujnovsky, I., Hirayama, J., Cermakian, N., Cato, A. C. and Sassone-Corsi, P. (2007) Light-inducible and clock-controlled expression of MAP kinase phosphatase 1 in mouse central pacemaker neurons. *Journal of biological rhythms*, **22**, 127-139.

[71] Shimizu, K., Kobayashi, Y., Nakatsuji, E., Yamazaki, M., Shimba, S., Sakimura, K. and Fukada, Y. (2016) SCOP/PHLPP1beta mediates circadian regulation of long-term recognition memory. *Nature communications*, **7**, 12926.

[72] Gallego, M. and Virshup, D. M. (2007) Post-translational modifications regulate the ticking of the circadian clock. Nature reviews. *Molecular cell biology*, **8**, 139-148.

[73] Corda, D. and Di Girolamo, M. (2003) Functional aspects of protein mono-ADP-ribosylation. *Embo J*, **22**, 1953-1958.

[74] Freiman, R. N. and Tjian, R. (2003) Regulating the regulators: lysine modifications make their mark. *Cell*, **112**, 11-17.

[75] Lowrey, P. L., Shimomura, K., Antoch, M. P., Yamazaki, S., Zemenides, P. D., Ralph, M. R., Menaker, M. and Takahashi, J. S. (2000) Positional syntenic cloning and functional characterization of the mammalian circadian mutation tau. *Science* (New York, N.Y.), **288**, 483-492.

[76] Toh, K. L., Jones, C. R., He, Y., Eide, E. J., Hinz, W. A., Virshup, D. M., Ptacek, L. J. and Fu, Y. H. (2001) An hPer2 phosphorylation site mutation in familial advanced sleep phase syndrome. *Science* (New York, N.Y.), **291**, 1040-1043.

[77] Ebisawa, T., Uchiyama, M., Kajimura, N., Mishima, K., Kamei, Y., Katoh, M., Watanabe, T., Sekimoto, M., Shibui, K., Kim, K. et al. (2001) Association of structural polymorphisms in the human period3 gene with delayed sleep phase syndrome. *EMBO reports*, **2**, 342-346.

[78] Akashi, M., Tsuchiya, Y., Yoshino, T. and Nishida, E. (2002) Control of intracellular dynamics of mammalian period proteins by casein kinase I epsilon (CKIepsilon) and CKIdelta in cultured cells. *Mol Cell Biol*, **22**, 1693-1703.

[79] Eide, E. J., Woolf, M. F., Kang, H., Woolf, P., Hurst, W., Camacho, F., Vielhaber, E. L., Giovanni, A. and Virshup, D. M. (2005) Control

of mammalian circadian rhythm by CKIepsilon-regulated proteasome-mediated PER2 degradation. *Mol Cell Biol*, **25**, 2795-2807.
[80] Lawler, S., Fleming, Y., Goedert, M. and Cohen, P. (1998) Synergistic activation of SAPK1/JNK1 by two MAP kinase kinases in vitro. *Curr Biol*, **8**, 1387-1390.
[81] Kishimoto, H., Nakagawa, K., Watanabe, T., Kitagawa, D., Momose, H., Seo, J., Nishitai, G., Shimizu, N., Ohata, S., Tanemura, S. et al. (2003) Different properties of SEK1 and MKK7 in dual phosphorylation of stress-induced activated protein kinase SAPK/JNK in embryonic stem cells. *The Journal of biological chemistry*, **278**, 16595-16601.
[82] Yamasaki, T., Kawasaki, H. and Nishina, H. (2012) Diverse Roles of JNK and MKK Pathways in the Brain. *Journal of signal transduction*, **2012**, 459265.
[83] Uchida, Y., Osaki, T., Yamasaki, T., Shimomura, T., Hata, S., Horikawa, K., Shibata, S., Todo, T., Hirayama, J. and Nishina, H. (2012) Involvement of stress kinase mitogen-activated protein kinase kinase 7 in regulation of mammalian circadian clock. *The Journal of biological chemistry*, **287**, 8318-8326.
[84] Yamasaki, T., Deki-Arima, N., Kaneko, A., Miyamura, N., Iwatsuki, M., Matsuoka, M., Fujimori-Tonou, N., Okamoto-Uchida, Y., Hirayama, J., Marth, J. D. et al. (2017) Age-dependent motor dysfunction due to neuron-specific disruption of stress-activated protein kinase MKK7. *Scientific reports*, **7**, 7348.
[85] Yoshitane, H., Honma, S., Imamura, K., Nakajima, H., Nishide, S. Y., Ono, D., Kiyota, H., Shinozaki, N., Matsuki, H., Wada, N. et al. (2012) JNK regulates the photic response of the mammalian circadian clock. *EMBO reports*, **13**, 455-461.
[86] Sanada, K., Harada, Y., Sakai, M., Todo, T. and Fukada, Y. (2004) Serine phosphorylation of mCRY1 and mCRY2 by mitogen-activated protein kinase. *Genes to cells : devoted to molecular & cellular mechanisms*, **9**, 697-708.
[87] Eide, E. J., Vielhaber, E. L., Hinz, W. A. and Virshup, D. M. (2002) The circadian regulatory proteins BMAL1 and cryptochromes are

substrates of casein kinase Iepsilon. *The Journal of biological chemistry*, **277**, 17248-17254.

[88] Sanada, K., Okano, T. and Fukada, Y. (2002) Mitogen-activated protein kinase phosphorylates and negatively regulates basic helix-loop-helix-PAS transcription factor BMAL1. *The Journal of biological chemistry*, **277**, 267-271.

[89] Husse, J., Eichele, G. and Oster, H. (2015) Synchronization of the mammalian circadian timing system: Light can control peripheral clocks independently of the SCN clock: alternate routes of entrainment optimize the alignment of the body's circadian clock network with external time. *BioEssays: news and reviews in molecular, cellular and developmental biology*, **37**, 1119-1128.

[90] Akashi, M. and Nishida, E. (2000) Involvement of the MAP kinase cascade in resetting of the mammalian circadian clock. *Genes Dev*, **14**, 645-649.

[91] Yamajuku, D., Inagaki, T., Haruma, T., Okubo, S., Kataoka, Y., Kobayashi, S., Ikegami, K., Laurent, T., Kojima, T., Noutomi, K. et al. (2012) Real-time monitoring in three-dimensional hepatocytes reveals that insulin acts as a synchronizer for liver clock. *Scientific reports*, **2**, 439.

[92] Sato, M., Murakami, M., Node, K., Matsumura, R. and Akashi, M. (2014) The role of the endocrine system in feeding-induced tissue-specific circadian entrainment. *Cell reports*, **8**, 393-401.

[93] Hayashi, Y., Sanada, K., Hirota, T., Shimizu, F. and Fukada, Y. (2003) p38 mitogen-activated protein kinase regulates oscillation of chick pineal circadian clock. *The Journal of biological chemistry*, **278**, 25166-25171.

[94] Kon, N., Sugiyama, Y., Yoshitane, H., Kameshita, I. and Fukada, Y. (2015) Cell-based inhibitor screening identifies multiple protein kinases important for circadian clock oscillations. *Communicative & integrative biology*, **8**, e982405.

[95] Pizzio, G. A., Hainich, E. C., Ferreyra, G. A., Coso, O. A. and Golombek, D. A. (2003) Circadian and photic regulation of ERK, JNK and p38 in the hamster SCN. *Neuroreport*, **14**, 1417-1419.

[96] Chansard, M., Molyneux, P., Nomura, K., Harrington, M. E. and Fukuhara, C. (2007) c-Jun N-terminal kinase inhibitor SP600125 modulates the period of mammalian circadian rhythms. *Neuroscience*, **145**, 812-823.

[97] Tsuchiya, Y., Minami, I., Kadotani, H., Todo, T. and Nishida, E. (2013) Circadian clock-controlled diurnal oscillation of Ras/ERK signaling in mouse liver. *Proceedings of the Japan Academy. Series B, Physical and biological sciences*, **89**, 59-65.

[98] Nishimoto, S. and Nishida, E. (2006) MAPK signalling: ERK5 versus ERK1/2. *EMBO reports*, **7**, 782-786.

[99] Doi, M., Hirayama, J. and Sassone-Corsi, P. (2006) Circadian regulator CLOCK is a histone acetyltransferase. *Cell*, **125**, 497-508.

Chapter 3

TOXICOLOGICAL SIGNIFICANCE OF MAPK ACTIVATION IN CADMIUM-INDUCED KIDNEY CELL DEATH

Masato Matsuoka[*]
Department of Hygiene and Public Health,
Division of Environmental and Occupational Medicine,
Tokyo Women's Medical University, Shinjuku, Tokyo, Japan

ABSTRACT

Cadmium, one of the toxic metals, is an important occupational and environmental pollutant that damages various organs, especially the kidney. Animal and cultured cell studies show that cadmium exposure induces apoptotic cell death in proximal tubular epithelial cells and glomerular mesangial cells. However, cadmium-induced cellular stress can activate the signaling pathways responsible for both apoptosis and anti-apoptosis. Accumulating evidence indicates that mitogen-activated protein kinases (MAPKs), including extracellular signal-regulated kinase 1/2 (ERK1/2), ERK5 (also known as big MAPK 1 [BMK1]), c-Jun NH$_2$-

---

[*] Corresponding Author's E-mail: masato.matsuoka@twmu.ac.jp.

terminal kinase (JNK), and p38, are phosphorylated and activated in kidney cells following exposure to cadmium. This short review summarizes the toxicological significance of MAPK activation in cadmium-induced cell death and survival in five types of kidney cells: renal tubular epithelial cells, glomerular mesangial cells, glomerular endothelial cells, podocyte cells, and human embryonic kidney cells. Most studies in kidney cells show that, generally, the activation of the JNK and p38 pathways leads to cell death while the activation of the ERK pathway leads to cell survival. However, the type of kidney cell and severity of cadmium-induced cellular stress appear to determine the effect of MAPKs on cell fate.

Keywords: cadmium, MAPKs, kidney cells, apoptotic cell death

1. Cadmium and Nephrotoxicity

The metallic element cadmium has been used in a large variety of industrial processes, such as electroplating manufacturing parts, coloring plastics and glass, and constructing nickel-cadmium batteries. The kidney is considered the critical target organ for cadmium toxicity in the general population, as well as occupationally exposed populations [1]. With chronic exposure, cadmium accumulates in the epithelial cells of the proximal tubule and results in generalized reabsorptive dysfunction [2]. The earliest manifestations of cadmium-induced renal damage are increased urinary excretion of low-molecular weight proteins such as β_2-microglobulin [3]. With progression, more generalized tubular dysfunction occurs with urinary losses of glucose, amino acids, bicarbonate, and phosphate [4]. Exposure to cadmium at higher levels results in a decline in the glomerular filtration rate and albuminuria [2-4].

2. Cadmium-Induced Kidney Cell Death

Cadmium has been shown to induce apoptosis (type I cell death) in the proximal tubules of experimental animals [5-7] and in cultured proximal tubular epithelial cells [8-11]. Although the proximal tubules are the primary target of cadmium-induced kidney damage, cadmium also induces apoptosis in mesangial cells [12-14]. Generally, in cell culture systems, cadmium at low and moderate concentrations (0.1 – 10 μM) primarily causes apoptosis; at high concentrations (> 50 μM), necrosis (type III cell death) becomes evident [15].

Although the pathophysiological significance of autophagy (type II cell death) in cadmium-induced cell death is still unclear [16], accumulating evidence indicates that cadmium causes autophagy in proximal tubular cells [11, 17] and mesangial cells [18], as well as in the kidneys of mouse [19, 20], rat [17], chicken [21], and common carp [22]. These findings are compatible with those showing that cadmium exposure induces endoplasmic reticulum (ER) stress in proximal tubular cells [11, 23-25].

3. ACTIVATION OF MAPKs IN KIDNEY CELLS EXPOSED TO CADMIUM

In LLC-PK$_1$, an established porcine renal epithelial cell line with characteristics of the proximal tubule [26], we found that cadmium exposure induces the expression of immediate early genes including c-*fos*, c-*jun*, c-*myc*, and *Egr-1* (also known as *zif/268* and *Krox-24*) [8]. Because the expression of c-*fos* and c-*jun* genes is regulated by mitogen-activated protein kinases (MAPKs), including extracellular signal-regulated kinase 1/2 (ERK1/2), c-Jun NH$_2$-terminal kinase (JNK)/stress-activated protein kinase, and p38 [27, 28], we examined the effects of cadmium exposure on these MAPK pathways in renal proximal tubular cells, LLC-PK$_1$ cells and HK-2, an immortalized proximal tubule epithelial cell line derived from normal adult human kidney [29].

Cadmium exposure increases the activity of JNK and the level of the c-Jun protein phosphorylated on Ser63 and Ser73 in LLC-PK$_1$ cells [30]. It also induces the phosphorylation of ERK1/2 (ERK2/p42 and ERK1/p44) on Thr202 and Tyr204, ERK5 (also known as big MAPK 1 [BMK1]) on Thr218 and Tyr220, JNK on Thr183 and Tyr185, and p38 on Thr180 and Tyr182 in LLC-PK$_1$ cells [31] or HK-2 cells [11, 32, 33]. In rat mesangial cells isolated from the glomeruli of kidneys, the elevation of ERK and JNK activities was also reported by other investigators [34, 35]. Since these reports, cadmium-induced MAPK activation in kidney cells has been well confirmed by a substantial number of toxicological studies.

In addition, other metals (or metalloids), including lead [36], mercury [37, 38], zinc [39], and arsenic [40], can induce the activation of the ERK, JNK, or p38 pathways in kidney cells. Generally, ERK is activated by mitogenic stimuli, and JNK and p38 are activated by cellular stresses such as ultraviolet radiation, ionizing radiation, heat shock, chemical mutagens, and cytokines [41]. Our experimental findings paired with these collectively show that MAPKs are also activated by environmental pollution metals (or metalloids) including cadmium in kidney cells. Notably, MAPK activation by cadmium exposure is not specific to kidney cells but is also observed in a variety of cell types, including neuronal cells [42-44], pulmonary cells [45], hepatic cells [46], testicular cells [47], bone cells [48], lymphocytes [49, 50], and fibroblasts [51].

4. ROLE OF MAPK ACTIVATION IN CADMIUM-INDUCED KIDNEY CELL DEATH

In general, the phosphorylation of JNK and p38 stimulates apoptotic cell death, whereas the phosphorylation of ERK promotes normal cell proliferation [52]. To investigate the functional roles of MAPK activation in cadmium-induced cell death or cell survival, the conventional inhibitors of each MAPK pathway are used. While both U0126 and PD98059 are MAPK/ERK kinase (MEK) inhibitors, U0126 is more potent than PD98059. U0126 is equally potent toward MEK1 and MEK2, and PD98059 is more potent toward MEK1 [53-56]. The MEK5 inhibitor BIX02189 blocks ERK5 phosphorylation without affecting the phosphorylation of ERK1/2 [57]. SB203580 and SP600125 are the most commonly used inhibitors of p38 and JNK, respectively [58, 59]. The dominant-negative form of JNK and siRNA against JNK1 are also used to suppress JNK activity and expression, respectively. So far, the toxicological significance of cadmium-induced MAPK activation has been investigated in five types of kidney cells, all of which are discussed below.

4.1. Renal Tubular Epithelial Cells

LLC-PK$_1$ and HK-2 are the representative renal proximal tubular cell lines that are used for the nephrotoxicological studies. In LLC-PK$_1$ cells exposed to cadmium, two waves of apoptosis are noted 6 and 48 h after cadmium (10 µM) removal from the apical compartment. The presence of p38 inhibitor (Calbiochem) during recovery in cadmium-treated cultures results in significant decrease in delayed apoptosis and active caspase-3 levels. By contrast, incubation with PD98059, an MEK inhibitor, fails to alter the apoptotic fraction but increases activated caspase-3 levels in cadmium-treated cells [60]. Pretreatment of LLC-PK$_1$ cells with SP600125, a JNK inhibitor, inhibits the cellular damage and apoptosis induced by cadmium exposure (20 µM for 4 h), as evaluated by phase-contrast microscopy and Hoechst staining, respectively [24]. Pretreatment with selenium, an antioxidant metalloid, partially suppresses the phosphorylation of JNK, caspase-3 activation, and apoptosis in LLC-PK$_1$ cells exposed to 40 µM cadmium for 12 h [61]. We found that cadmium exposure causes ER stress in LLC-PK$_1$ cells and HK-2 cells [11, 23]. In HK-2 cells, treatment with salubrinal, an ER stress inhibitor, suppresses the phosphorylation of JNK and p38, not ERK1/2, protecting the cells from apoptosis induced by 24-h exposure to 20 µM cadmium [11]. These results imply that the pathways of JNK and p38, not ERK1/2, stimulate apoptotic cell death in these renal proximal tubular cell lines.

Similar to ERK1/2, ERK5 is rapidly and transiently phosphorylated after cadmium exposure [33]; this kinase is implicated in cell survival, differentiation, cell proliferation, and anti-apoptotic signaling [62]. Pretreatment of HK-2 cells with BIX02189, an inhibitor of MEK5, suppresses the phosphorylation of ERK5, not ERK1/2, at 2 h of exposure to

50 μM cadmium. At 4 h, this pretreatment suppresses the accumulation of phosphorylated cAMP response element-binding protein (CREB), activating transcription factor-1 (ATF-1), and mobility-shifted c-Fos protein. BIX02189 treatment enhances the activation of caspase-3 and apoptotic cell death in HK-2 cells exposed to 20 μM cadmium for 16 h, suggesting that the ERK5 pathway exerts an anti-apoptotic role by activating transcription factors associated with cell survival in HK-2 cells [33].

The pathways of JNK and p38 were also reportedly involved in apoptosis or parthanatos (poly[ADP-ribose] polymerase-1 [PARP-1]-dependent cell death) induced by cadmium exposure (5 μM for 12 h) in primary rat proximal tubular cells treated with SP600125 and SB203580, respectively [63, 64]. However, the inhibition of the ERK1/2 pathway with U0126 increases cadmium-induced apoptosis [64]. These findings also indicate the JNK and p38 pathways promote death signals and the ERK1/2 pathway mediates survival signals in primary cultured proximal tubular cells exposed to cadmium.

Consistent with the above findings, inhibitor studies showed the pro-apoptotic functional role of JNK and p38 in apoptosis induced by cadmium exposure (5 μM for 12 or 24 h) in NRK-52E rat renal tubular epithelial cells [64] and mouse renal tubular epithelial cells [65]. Furthermore, pro-apoptotic function of JNK was confirmed by demonstrating that the dominant-negative mutant of JNK can suppress apoptosis in NRK-52E cells exposed to 10 μM cadmium for 60 – 72 h [66]. However, in experiments using U0126, the ERK1/2 pathway was shown to suppress ER stress–mediated apoptosis by activating autophagy in NRK-52E cells [64].

4.2. Renal Glomerular Mesangial Cells

In primary mesangial cells isolated from the kidneys of adult mice, exposure to 10 μM cadmium for 0.5 – 6 h induces the biphasic phosphorylation of ERK1/2 with peaks at 1 h and 6 h. Exposure to 10 μM and 20 μM cadmium for 6 h induces apoptosis in 24% and 56% of mesangial cells, respectively. The inhibition of Ca^{2+}/calmodulin-dependent protein kinase II (CaMK-II) with methoxybenzenesulfonamide (KN93) suppresses cadmium-induced apoptosis and restores cell viability. However, the MEK inhibitor PD98059 fails to show any protective effects [12]. Cadmium exposure (10 – 40 µM for 6 h) also induces the phosphorylation of JNK and p38 in quiescent mouse mesangial cells. The inhibition of p38 with SB203580, but not that of JNK with SP600125, increases cell viability by suppressing apoptosis (annexin-V–positive and propidium iodide–negative cells) in preference to apoptosis-like cell death (annexin-V–positive and propidium iodide–positive cells) following exposure to 20 µM cadmium for 6 h [13]. Treatment with PD98059 increases caspase-3 cleavage product in primary rat mesangial cells exposed to 0.5 M cadmium for 24 h, indicating the anti-apoptotic property of ERK1/2 activation [14]. Collectively, these findings indicate that cadmium induces apoptosis in mesangial cells with a CaMK-II- and p38-dependent but ERK- and JNK-independent manner.

In contrast to ERK-independent apoptosis in primary mesangial cells, ERK signaling was reported to be responsible for cadmium-induced cell death in immortalized mouse renal mesangial SV40-Mes13 cells, which are derived from SV40 transgenic mice [67, 68]. Suppression of autophagy with 3-methyladenine increases cell viability in SV40-Mes13 cells exposed to 6 µM cadmium for 28 h. However, the inhibition of ERK phosphorylation by PD98059 and U0126 suppresses autophagy induced by cadmium exposure (6 µM for 28 h), and treatment with BAPTA-AM, a cell permeable free calcium chelator, eliminates ERK activation induced by cadmium exposure (6 µM for 1 h) [18]. Based on these results using SV40-Mes13 cells, it is postulated that the cadmium-induced Ca^{2+}-ERK signaling pathway leads predominantly to autophagic cell death and in a lesser degree to apoptotic cell death in mesangial cells [69].

Exposure of primary mesangial cells isolated from human renal tissue to 4 µM cadmium for 24 h induces a decrease in proliferation and an increase in apoptosis. However, with pretreatment with SP600125, only the decrease in proliferation is reversed, suggesting that the JNK pathway mediates the inhibitory effects of cadmium on mesangial cell proliferation [70]. In contrast to the p38 pathway, the JNK pathway does not significantly contribute to cadmium-induced cell death in mesangial cells.

4.3. Renal Glomerular Endothelial Cells

Little is known about the functional effects of cadmium exposure on the glomerular endothelium, which is a critical part of the glomerular filtration barrier [71]. Exposure to 1 μM cadmium for 1 h increases permeability in primary human renal glomerular endothelial cells (HRGECs) without cytotoxic effects. While cadmium exposure induces the phosphorylation of ERK1/2, JNK, and p38, treatment with SB203580 but not PD98059 and SP600125 partially prevents cadmium-induced hyperpermeability in HRGECs, implying the involvement of p38 activation in the dysfunction of glomerular filtration [72]. In addition, exposure of HRGECs to 4 μM cadmium for 24 h fails to induce apoptosis but activates the nuclear factor-κB (NF-κB) pathway. Treatment with SP600125 partially reverses apoptosis induced by cadmium exposure (4 μM for 24 h) in HRGECs pretreated with pyrrolidine dithiocarbamate (PDTC), a potent NF-κB inhibitor. It is suggested that the NF-κB pathway prevents cell death partially by inhibiting the JNK pathway [73].

4.4. Renal Glomerular Podocyte Cells

Renal podocytes are highly specialized epithelial cells that line the urinary surface of the glomerular capillary tuft and maintain glomerular filtration [74, 75]. Exposure of human renal podocytes to 4 μM cadmium for 1 – 24 h activates the JNK pathway and induces the expression of c-Fos and c-Jun proteins but does not significantly affect cell proliferation, viability, and apoptosis at 24 h exposure [75]. To date, the role of activation of JNK and other MAPKs by cadmium exposure in renal podocytes remains clear.

4.5. Human Embryonic Kidney Cells

Although human embryonic kidney (HEK) 293 cells are originally isolated from primary human embryo kidney cells transformed by sheared adenovirus 5 DNA, HEK293 cells have properties of neuronal lineage cells and are not typical kidney cells [76, 77]. Therefore, the findings concerning the role of cadmium-induced MAPK activation in HEK293 cells cannot be interpreted as those in kidney cells.

Cotreatment with U0126 protects HEK293 cells from cellular damage induced by cadmium exposure (1 µM for 48 h) [78]. Treatment with manganese, another metal, inhibits sustained phosphorylation of ERK and cytotoxicity induced by cadmium exposure (2 µM for 24 h) in HEK293 cells [79]. Similarly, treatment with phorbol ester, an activator of protein kinase C, protects HEK293 cells against toxicity induced by 24-h exposure to 0.5 and 1 µM cadmium and diminishes ERK phosphorylation induced by 24-h exposure to 2 µM cadmium [80]. These findings suggest that ERK activation exceptionally stimulates the death signal pathway in HEK293 cells exposed to cadmium. The involvement of the ERK pathway in the cadmium-induced apoptotic cell death is also observed in certain cell types, such as lymphocytes [50], hepatocytes [81, 82], and neuronal cells [83, 84]. This noncanonical association between ERK1/2 and cadmium toxicity is well reviewed elsewhere [15, 85, 86].

The knockdown of JNK1 expression with siRNA suppresses the reduction of cell viability in HEK293 cells exposed to 30 – 60 µM cadmium for 48 h, indicating the involvement of the JNK1 pathway in cadmium cytotoxicity [87]. However, at 12 h of exposure, cadmium stimulates HEK293 cell proliferation at low concentrations (0.05 and 0.5 µM) but induces apoptosis at high concentrations (50 and 500 µM). In addition, treatment with PD98059 and SP600125 suppresses cell proliferation induced by cadmium exposure (0.5 µM for 12 h) in HEK293 cells [38]. Thus, the cell fate and the functional role of MAPK activation appear to be greatly different depending on the exposure dose of cadmium (i.e., the severity of cellular stress).

CONCLUSION

The well-known nephrotoxic metal, cadmium, induces activation of several MAPK signaling pathways, including ERK1/2, ERK5, JNK, and p38, in various types of kidney cells. *In vitro* experiments using pharmacological inhibitors such as U0126 and PD98059 (ERK1/2 pathway inhibitors), BIX02189 (ERK5 pathway inhibitor), SP600125 (JNK pathway inhibitor), and SB203580 (p38 pathway inhibitor) show that, generally, JNK and p38 transmit pro-apoptotic signaling and ERK transmits anti-apoptotic signaling in renal tubular epithelial cells. In contrast to the p38 pathway, the JNK pathway does not significantly contribute to cadmium-induced apoptosis in glomerular mesangial cells. However, the ERK pathway may be involved in the autophagic cell death of immortalized mouse renal mesangial cells (SV40-Mes13) and apoptotic cell death of human embryonic kidney cells (HEK293) induced by cadmium exposure. The type of kidney cell and the severity of cadmium-induced cellular stress appear to influence the effect of MAPKs on cell fate determination. The possible involvement of MAPKs in the cadmium-induced ER stress–mediated cell death and induction of autophagy in kidney cells must be clarified by future studies. More importantly, *in vivo* experiments using animal models will be needed to reveal the contribution of MAPK activation, especially that of ERK pathway, to cadmium-induced kidney damage and subsequent recovery.

REFERENCES

[1] International Programme on Chemical Safety. (1992). *Environmental Health Criteria* 134. Cadmium. Geneva: World Health Organization.

[2] Prozialeck, W. C. & Edwards, J. R. (2012). "Mechanisms of cadmium-induced proximal tubule injury: New insights with implications for biomonitoring and therapeutic interventions." *Journal of Pharmacology and Experimental Therapeutics, 343*, 2-12.

[3] Nordberg, G. F., Nogawa, K., Nordberg, M. & Friberg, L. T. (2007). "Cadmium." In *Handbook on the Toxicology of Metals*, edited by Nordberg, G. F., Fowler, B. A., Nordberg, M., and Friberg, L. T., 445-86. Burlington: Academic Press.

[4] Johri, N., Jacquillet, G. & Unwin, R. (2010). "Heavy metal poisoning: the effects of cadmium on the kidney." *BioMetals, 23*, 783-92.

[5] Hamada, T., Nakano, S., Iwai, S., Tanimoto, A., Ariyoshi, K. & Koide, O. (1991). "Pathological study on beagles after long-term oral administration of cadmium." *Toxicologic Pathology, 19*, 138-47.

[6] Tanimoto, A., Hamada, T. & Koide, O. (1993). "Cell death and regeneration of renal proximal tubular cells in rats with subchronic cadmium intoxication." *Toxicologic Pathology, 21*, 341-52.

[7] Hamada, T., Tanimoto, A. & Sasaguri, Y. (1997). "Apoptosis induced by cadmium." *Apoptosis, 2*, 359-67.

[8] Matsuoka, M. & Call, K. M. (1995). "Cadmium-induced expression of immediate early genes in LLC-PK$_1$ cells." *Kidney International, 48*, 383-9.

[9] Lee, W. K., Torchalski, B. & Thévenod, F. (2007). "Cadmium-induced ceramide formation triggers calpain-dependent apoptosis in cultured kidney proximal tubule cells." *American Journal of Physiology Cell Physiology, 293*, C839-47.

[10] Lee, W. K. & Thévenod, F. (2008). "Novel roles for ceramides, calpains and caspases in kidney proximal tubule cell apoptosis: Lessons from *in vitro* cadmium toxicity studies." *Biochemical Pharmacology, 76*, 1323-32.

[11] Komoike, Y., Inamura, H. & Matsuoka, M. (2012). "Effects of salubrinal on cadmium-induced apoptosis in HK-2 human renal proximal tubular cells." *Archives of Toxicology, 86*, 37-44.

[12] Liu, Y. & Templeton, D. M. (2007). "Cadmium activates CaMK-II and initiates CaMK-II-dependent apoptosis in mesangial cells." *FEBS Letters, 581*, 1481-6.

[13] Liu, Y. & Templeton, D. M. (2008). "Initiation of caspase-independent death in mouse mesangial cells by Cd^{2+}: Involvement of p38 kinase and CaMK-II." *Journal of Cellular Physiology, 217*, 307-18.

[14] Xiao, W., Liu, Y. & Templeton, D. M. (2009). "Pleiotropic effects of cadmium in mesangial cells." *Toxicology and Applied Pharmacology, 238*, 315-26.

[15] Templeton, D. M. & Liu, Y. (2010). "Multiple roles of cadmium in cell death and survival." *Chemico-Biological Interactions, 188*, 267-75.

[16] Thévenod, F. & Lee, W. K. (2015). "Live and let die: Roles of autophagy in cadmium nephrotoxicity." *Toxics, 3*, 130-51. doi:10.3390/toxics3020130.

[17] Chargui, A., Zekri, S., Jacquillet, G., Rubera, I., Ilie, M., Belaid, A., Duranton, C., Tauc, M., Hofman, P., Poujeol, P., El May, M. V. & Mograbi, B. (2011). "Cadmium-induced autophagy in rat kidney: An early biomarker of subtoxic exposure." *Toxicological Sciences, 121*, 31-42.

[18] Wang, S. H., Shih, Y. L., Ko, W. C., Wei, Y. H. & Shih, C. M. (2008). "Cadmium-induced autophagy and apoptosis are mediated by a calcium signaling pathway." *Cellular and Molecular Life Sciences, 65*, 3640-52.

[19] Luo, B., Lin, Y., Jiang, S., Huang, L., Yao, H., Zhuang, Q., Zhao, R., Liu, H., He, C. & Lin, Z. (2016). "Endoplasmic reticulum stress eIF2α–ATF4 pathway-mediated cyclooxygenase-2 induction regulates cadmium-induced autophagy in kidney." *Cell Death & Disease, 7*, e2251. doi:10.1038/cddis.2016.78.

[20] Yuan, Y., Ma, S., Qi, Y., Wei, X., Cai, H., Dong, L., Lu, Y., Zhang, Y. & Guo, Q. (2016). "Quercetin inhibited cadmium-induced autophagy in the mouse kidney via inhibition of oxidative stress." *Journal of Toxicologic Pathology, 29*, 247-52.

[21] Shi, Q., Jin, X., Fan, R., Xing, M., Guo, J., Zhang, Z., Zhang, J. & Xu, S. (2019). "Cadmium-mediated miR-30a-GRP78 leads to JNK-dependent autophagy in chicken kidney." *Chemosphere, 215*, 710-5.

[22] Gao, D., Xu, Z., Kuang, X., Qiao, P., Liu, S., Zhang, L., He, P., Jadwiga, W. S., Wang, Y. & Min, W. (2014). "Molecular characterization and expression analysis of the autophagic gene Beclin 1 from the purse red common carp (*Cyprinus carpio*) exposed to

cadmium." *Comparative Biochemistry and Physiology Part C Toxicology & Pharmacology, 160,* 15-22.
[23] Liu, F., Inageda, K., Nishitai, G. & Matsuoka, M. (2006). "Cadmium induces the expression of Grp78, an endoplasmic reticulum molecular chaperone, in LLC-PK1 renal epithelial cells." *Environmental Health Perspectives, 114,* 859-64.
[24] Yokouchi, M., Hiramatsu, N., Hayakawa, K., Kasai, A., Takano, Y., Yao, J. & Kitamura, M. (2007). "Atypical, bidirectional regulation of cadmium-induced apoptosis via distinct signaling of unfolded protein response." *Cell Death & Differentiation, 14,* 1467-74.
[25] Yokouchi, M., Hiramatsu, N., Hayakawa, K., Okamura, M., Du, S., Kasai, A., Takano, Y., Shitamura, A., Shimada, T., Yao, J. & Kitamura, M. (2008). "Involvement of selective reactive oxygen species upstream of proapoptotic branches of unfolded protein response." *Journal of Biological Chemistry, 283,* 4252-60.
[26] Gstraunthaler, G., Pfaller, W. & Kotanko, P. (1985). "Biochemical characterization of renal epithelial cell cultures (LLC-PK$_1$ and MDCK)." *American Journal of Physiology Renal Physiology, 248,* F536-44.
[27] Karin, M. (1995). "The regulation of AP-1 activity by mitogen-activated protein kinases." *Journal of Biological Chemistry, 270,* 16483-6.
[28] Whitmarsh, A. J. & Davis, R. J. (1996). "Transcription factor AP-1 regulation by mitogen-activated protein kinase signal transduction pathways." *Journal of Molecular Medicine, 74,* 589-607.
[29] Ryan, M. J., Johnson, G., Kirk, J., Fuerstenberg, S. M., Zager, R. A. & Torok-Storb, B. (1994). "HK-2: An immortalized proximal tubule epithelial cell line from normal adult human kidney." *Kidney International, 45,* 48-57.
[30] Matsuoka, M. & Igisu, H. (1998). "Activation of c-Jun NH$_2$-terminal kinase (JNK/SAPK) in LLC-PK$_1$ cells by cadmium." *Biochemical and Biophysical Research Communications, 251,* 527-32.
[31] Nakagawa, J., Nishitai, G., Inageda, K. & Matsuoka, M. (2007). "Phosphorylation of Stats at Ser727 in renal proximal tubular

epithelial cells exposed to cadmium." *Environmental Toxicology and Pharmacology, 24,* 252-9.

[32] Iwatsuki, M., Inageda, K. & Matsuoka, M. (2011). "Cadmium induces phosphorylation and stabilization of c-Fos in HK-2 renal proximal tubular cells." *Toxicology and Applied Pharmacology, 251,* 209-16.

[33] Kondo, M., Inamura, H., Matsumura, K. & Matsuoka, M. (2012). "Cadmium activates extracellular signal-regulated kinase 5 in HK-2 human renal proximal tubular cells." *Biochemical and Biophysical Research Communications, 421,* 490-3.

[34] Wang, Z. & Templeton, D. M. (1998). "Induction of c-*fos* proto-oncogene in mesangial cells by cadmium." *Journal of Biological Chemistry, 273,* 73-9.

[35] Ding, W. & Templeton, D. M. (2000). "Activation of parallel mitogen-activated protein kinase cascades and induction of c-*fos* by cadmium." *Toxicology and Applied Pharmacology, 162,* 93-9.

[36] Liu, C. M., Sun, Y. Z., Sun, J. M., Ma, J. Q. & Cheng, C. (2012). "Protective role of quercetin against lead-induced inflammatory response in rat kidney through the ROS-mediated MAPKs and NF-κB pathway." *Biochimica et Biophysica Acta, 1820,* 1693-703.

[37] Turney, K. D., Parrish, A. R., Orozco, J. & Gandolfi, A. J. (1999). "Selective activation in the MAPK pathway by Hg(II) in precision-cut rabbit renal cortical slices." *Toxicology and Applied Pharmacology, 160,* 262-70.

[38] Hao, C., Hao, W., Wei, X., Xing, L., Jiang, J. & Shang, L. (2009). "The role of MAPK in the biphasic dose-response phenomenon induced by cadmium and mercury in HEK293 cells." *Toxicology in Vitro, 23,* 660-6.

[39] Matsunaga, Y., Kawai, Y., Kohda, Y. & Gemba, M. (2005). "Involvement of activation of NADPH oxidase and extracellular signal-regulated kinase (ERK) in renal cell injury induced by zinc." *Journal of Toxicological Sciences, 30,* 135-44.

[40] Gong, X., Ivanov, V. N. & Hei, T. K. (2016). "2,3,5,6-Tetramethylpyrazine (TMP) down-regulated arsenic-induced heme oxygenase-1 and ARS2 expression by inhibiting Nrf2, NF-κB, AP-1

and MAPK pathways in human proximal tubular cells." *Archives of Toxicology*, *90*, 2187-200.
[41] Kyriakis, J. M. & Avruch, J. (1996). "Sounding the alarm: Protein kinase cascades activated by stress and inflammation." *Journal of Biological Chemistry*, *271*, 24313-6.
[42] Rockwell, P., Martinez, J., Papa, L. & Gomes, E. (2004). "Redox regulates COX-2 upregulation and cell death in the neuronal response to cadmium." *Cellular Signalling*, *16*, 343-53.
[43] Kim, S. D., Moon, C. K., Eun, S. Y., Ryu, P. D. & Jo, S. A. (2005). "Identification of ASK1, MKK4, JNK, c-Jun, and caspase-3 as a signaling cascade involved in cadmium-induced neuronal cell apoptosis." *Biochemical and Biophysical Research Communications*, *328*, 326-34.
[44] Chen, L., Liu, L., Luo, Y. & Huang, S. (2008). "MAPK and mTOR pathways are involved in cadmium-induced neuronal apoptosis." *Journal of Neurochemistry*, *105*, 251-61.
[45] Chuang, S. M., Wang, I. C. & Yang, J. L. (2000). "Roles of JNK, p38 and ERK mitogen-activated protein kinases in the growth inhibition and apoptosis induced by cadmium." *Carcinogenesis*, *21*, 1423-32.
[46] Souza, V., Escobar M. C., Bucio, L., Hernández, E., Gómez-Quiroz, L. E. & Gutiérrez Ruiz, M. C. (2009). "NADPH oxidase and ERK1/2 are involved in cadmium induced-STAT3 activation in HepG2 cells." *Toxicology Letters*, *187*, 180-6.
[47] Siu, E. R., Mruk, D. D., Porto, C. S. & Cheng, C. Y. (2009). "Cadmium-induced testicular injury." *Toxicology and Applied Pharmacology*, *238*, 240-9.
[48] Brama, M., Politi, L., Santini, P., Migliaccio, S. & Scandurra, R. (2012). "Cadmium-induced apoptosis and necrosis in human osteoblasts: Role of caspases and mitogen-activated protein kinases pathways." *Journal of Endocrinological Investigation*, *35*, 198-208.
[49] Galán, A., Garcia-Bermejo, M. L., Troyano, A., Vilaboa, N. E., de Blas, E., Kazanietz, M. G. & Aller, P. (2000). "Stimulation of p38 mitogen-activated protein kinase is an early regulatory event for the

cadmium-induced apoptosis in human promonocytic cells." *Journal of Biological Chemistry*, 275, 11418-24.

[50] Iryo, Y., Matsuoka, M., Wispriyono, B., Sugiura, T. & Igisu, H. (2000). "Involvement of the extracellular signal-regulated protein kinase (ERK) pathway in the induction of apoptosis by cadmium chloride in CCRF-CEM cells." *Biochemical Pharmacology*, 60, 1875-82.

[51] Sugisawa, N., Matsuoka, M., Okuno, T. & Igisu, H. (2004). "Suppression of cadmium-induced JNK/p38 activation and HSP70 family gene expression by LL-Z1640-2 in NIH3T3 cells." *Toxicology and Applied Pharmacology*, 196, 206-14.

[52] Wada, T. & Penninger, J. M. (2004). "Mitogen-activated protein kinases in apoptosis regulation." *Oncogene*, 23, 2838-49.

[53] Alessi, D. R., Cuenda, A., Cohen, P., Dudley, D. T. & Saltiel, A. R. (1995). "PD 098059 is a specific inhibitor of the activation of mitogen-activated protein kinase kinase in vitro and in vivo." *Journal of Biological Chemistry*, 270, 27489-94.

[54] Dudley, D. T., Pang, L., Decker, S. J., Bridges, A. J. & Saltiel, A. R. (1995). "A synthetic inhibitor of the mitogen-activated protein kinase cascade." *Proceedings of the National Academy of Sciences of the United States of America*, 92, 7686-9.

[55] Favata, M. F., Horiuchi, K. Y., Manos, E. J., Daulerio, A. J., Stradley, D. A., Feeser, W. S., Van Dyk, D. E., Pitts, W. J., Earl, R. A., Hobbs, F., Copeland, R. A., Magolda, R. L., Scherle, P. A. & Trzaskos, J. M. (1998). "Identification of a novel inhibitor of mitogen-activated protein kinase kinase." *Journal of Biological Chemistry*, 273, 18623-32.

[56] Schonhoff, C. M., Bulseco, D. A., Brancho, D. M., Parada, L. F. & Ross, A. H. (2001). "The Ras–ERK pathway is required for the induction of neuronal nitric oxide synthase in differentiating PC12 cells." *Journal of Neurochemistry*, 78, 631-9.

[57] Tatake, R. J., O'Neill, M. M., Kennedy, C. A., Wayne, A. L., Jakes, S., Wu, D., Kugler, Jr. S. Z., Kashem, M. A., Kaplita, P. & Snow, R. J. (2008). "Identification of pharmacological inhibitors of the

MEK5/ERK5 pathway." *Biochemical and Biophysical Research Communications*, *377*, 120-5.

[58] Cuenda, A., Rouse, J., Doza, Y. N., Meier, R., Cohen, P., Gallagher, T. F., Young, P. R. & Lee, J. C. (1995). "SB 203580 is a specific inhibitor of a MAP kinase homologue which is stimulated by cellular stresses and interleukin-1." *FEBS Letters*, *364*, 229-33.

[59] Bennett, B. L., Sasaki, D. T., Murray, B. W., O'Leary, E. C., Sakata, S. T., Xu, W., Leisten, J. C., Motiwala, A., Pierce, S., Satoh, Y., Bhagwat, S. S., Manning, A. M. & Anderson, D. W. (2001). "SP600125, an anthrapyrazolone inhibitor of Jun N-terminal kinase." *Proceedings of the National Academy of Sciences of the United States of America*, *98*, 13681-6.

[60] Stinson, L. J., Darmon, A. J., Dagnino, L. & D'Souza, S. J. A. (2003). "Delayed apoptosis post-cadmium injury in renal proximal tubule epithelial cells." *American Journal of Nephrology*, *23*, 27-37.

[61] Liu, Y., Zhang, S. P. & Cai, Y. Q. (2007). "Cytoprotective effects of selenium on cadmium-induced LLC-PK$_1$ cells apoptosis by activating JNK pathway." *Toxicology in Vitro*, *21*, 677-84.

[62] Drew, B. A., Burow, M. E. & Beckman, B. S. (2012). "MEK5/ERK5 pathway: The first fifteen years." *Biochimica et Biophysica Acta*, *1825*, 37-48.

[63] Luo, T., Yuan, Y., Yu, Q., Liu, G., Long, M., Zhang, K., Bian, J., Gu, J., Zou, H., Wang, Y., Zhu, J., Liu, X. & Liu, Z. (2017). "PARP-1 overexpression contributes to Cadmium-induced death in rat proximal tubular cells via parthanatos and the MAPK signalling pathway." *Scientific Reports*, *7*, 4331. doi:10.1038/s41598-017-04555-2.

[64] Luo, T., Zhang, H., Yu, Q., Liu, G., Long, M., Zhang, K., Liu, W., Song, R., Bian, J., Gu, J., Zou, H., Liu, X., Yuan, Y. & Liu, Z. (2018). "ERK1/2 MAPK promotes autophagy to suppress ER stress-mediated apoptosis induced by cadmium in rat proximal tubular cells." *Toxicology in Vitro*, *52*, 60-9.

[65] Gu, J., Dai, S., Liu, Y., Liu, H., Zhang, Y., Ji, X., Yu, F., Zhou, Y., Chen, L., Tse, W. K. F., Wong, C. K. C., Chen, B. & Shi, H. (2018). "Activation of Ca^{2+}-sensing receptor as a protective pathway to reduce

Cadmium-induced cytotoxicity in renal proximal tubular cells." *Scientific Reports*, *8*, 1092. doi:10.1038/s41598-018-19327-9.

[66] Kato, H., Katoh, R. & Kitamura, M. (2013). "Dual regulation of cadmium-induced apoptosis by mTORC1 through selective induction of IRE1 branches in unfolded protein response." *PLoS One*, *8*, e64344. doi:10.1371/journal.pone.0064344.

[67] MacKay, K., Striker, L. J., Stauffer, J. W., Agodoa, L. Y. & Striker, G. E. (1990). "Relationship of glomerular hypertrophy and sclerosis: Studies in SV40 transgenic mice." *Kidney International*, *37*, 741-8.

[68] Lipkowitz, M. S., Hanss, B., Tulchin, N., Wilson, P. D., Langer, J. C., Ross, M. D., Kurtzman, G. J., Klotman, P. E. & Klotman, M. E. (1999). "Transduction of renal cells *in vitro* and *in vivo* by adeno-associated virus gene therapy vectors." *Journal of the American Society of Nephrology*, *10*, 1908-15.

[69] Yang, L. Y., Wu, K. H., Chiu, W. T., Wang, S. H. & Shih, C. M. (2009). "The cadmium-induced death of mesangial cells results in nephrotoxicity." *Autophagy*, *5*, 571-2.

[70] Chen, X., Li, J., Cheng, Z., Xu, Y., Wang, X., Li, X., Xu, D., Kapron, C. M. & Liu, J. (2016). "Low dose cadmium inhibits proliferation of human renal mesangial cells via activation of the JNK pathway." *International Journal of Environmental Research and Public Health*, *13*, 990. doi:10.3390/ijerph13100990.

[71] Obeidat, M., Obeidat, M. & Ballermann, B. J. (2012). "Glomerular endothelium: A porous sieve and formidable barrier." *Experimental Cell Research*, *318*, 964-72.

[72] Li, L., Dong, F., Xu, D., Du, L., Yan, S., Hu, H., Lobe, C. G., Yi, F., Kapron, C. M. & Liu, J. (2016). "Short-term, low-dose cadmium exposure induces hyperpermeability in human renal glomerular endothelial cells." *Journal of Applied Toxicology*, *36*, 257-65.

[73] Zhang, H., Li, L., Wang, Y., Dong, F., Chen, X., Liu, F., Xu, D., Yi, F., Kapron, C. M. & Liu, J. (2016). "NF-κB signaling maintains the survival of cadmium-exposed human renal glomerular endothelial cells." *International Journal of Molecular Medicine*, *38*, 417-22.

[74] Tharaux, P. L. & Huber, T. B. (2012). "How many ways can a podocyte die?" *Seminars in Nephrology, 32*, 394-404.

[75] Chen, X., Xu, Y., Cheng, Z., Su, H., Liu, X., Xu, D., Kapron, C. & Liu, J. (2018). "Low-dose cadmium activates the JNK signaling pathway in human renal podocytes." *International Journal of Molecular Medicine, 41*, 2359-65.

[76] Shaw, G., Morse, S., Ararat, M. & Graham, F. L. (2002). "Preferential transformation of human neuronal cells by human adenoviruses and the origin of HEK 293 cells." *FASEB Journal, 16*, 869-71.

[77] Toth, K. & Wold, W. S. M. (2002). "HEK? No!" *Molecular Therapy, 5*, 654.

[78] Martin, P., Poggi, M. C., Chambard, J. C., Boulukos, K. E. & Pognonec, P. (2006). "Low dose cadmium poisoning results in sustained ERK phosphorylation and caspase activation." *Biochemical and Biophysical Research Communications, 350*, 803-7.

[79] Martin, P., Fareh, M., Poggi, M. C., Boulukos, K. E. & Pognonec, P. (2006). "Manganese is highly effective in protecting cells from cadmium intoxication." *Biochemical and Biophysical Research Communications, 351*, 294-9.

[80] Martin, P., Boulukos, K. E., Poggi, M. C. & Pognonec, P. (2009). "Long-term extracellular signal-related kinase activation following cadmium intoxication is negatively regulated by a protein kinase C-dependent pathway affecting cadmium transport." *FEBS Journal, 276*, 1667-79.

[81] Yiran, Z., Chenyang, J., Jiajing, W., Yan, Y., Jianhong, G., Jianchun, B., Xuezhong, L. & Zongping, L. (2013). "Oxidative stress and mitogen-activated protein kinase pathways involved in cadmium-induced BRL 3A cell apoptosis." *Oxidative Medicine and Cellular Longevity, 2013*, 516051. doi:10.1155/2013/516051.

[82] Zou, H., Liu, X., Han, T., Hu, D., Wang, Y., Yuan, Y., Gu, J., Bian, J., Zhu, J. & Liu, Z. P. (2015). "Salidroside protects against cadmium-induced hepatotoxicity in rats via GJIC and MAPK pathways." *PLoS One, 10*, e0129788. doi:10.1371/journal.pone. 0129788.

[83] Yuan, Y., Jiang, C., Hu, F., Wang, Q., Zhang, K., Wang, Y., Gu, J., Liu, X., Bian, J. & Liu, Z. (2015). "The role of mitogen-activated protein kinase in cadmium-induced primary rat cerebral cortical neurons apoptosis via a mitochondrial apoptotic pathway." *Journal of Trace Elements in Medicine and Biology, 29,* 275-83.

[84] Xu, C., Wang, X., Zhu, Y., Dong, X., Liu, C., Zhang, H., Liu, L., Huang, S. & Chen, L. (2016). "Rapamycin ameliorates cadmium-induced activation of MAPK pathway and neuronal apoptosis by preventing mitochondrial ROS inactivation of PP2A." *Neuropharmacology, 105,* 270-84.

[85] Martin, P. & Pognonec, P. (2010). "ERK and cell death: cadmium toxicity, sustained ERK activation and cell death." *FEBS Journal, 277,* 39-46.

[86] Thévenod, F. & Lee, W. K. (2013). "Cadmium and cellular signaling cascades: interactions between cell death and survival pathways." *Archives of Toxicology, 87,* 1743-86.

[87] Du, K., Takahashi, T., Kuge, S., Naganuma, A. & Hwang, G. W. (2014). "FBXO6 attenuates cadmium toxicity in HEK293 cells by inhibiting ER stress and JNK activation." *Journal of Toxicological Sciences, 39,* 861-6.

BIBLIOGRAPHY

Dermatology research focus on acne, melanoma and psoriasis

LCCN	2009037462
Type of material	Book
Main title	Dermatology research focus on acne, melanoma and psoriasis / David E. Roth, editor.
Published/Created	New York: Nova Science Publishers, c2010.
Description	xv, 332 p.: ill. (some col.); 27 cm.
ISBN	9781608760756 (hardcover)
	1608760758 (hardcover)
LC classification	RL131 .D47 2010
Related names	Roth, David E., 1963-
Contents	Acne: causes, treatment and myths / K.L.E. Hon, Alexander K.C. Leung -- Female acne vulgaris: a common skin disease affecting the relational quality of life: beneficial effects of hormonal contraceptives / Sabatini Rosa -- Shedding light on acne: from myth to science (laser and light therapy for acne) / Seung Yoon Celine Lee -- Acne: causes, treatment and myths / Neelam Muizzuddin -- Emerging trends in the diagnosis and management of atopic dermatitis / Mohamed

L. Elsaie, Sonal Choudhary -- Goeckerman regimen: genotoxic, pro-apoptotic, anti-inflammatory and anti-angiogenic effect / L. Borska ... [et al.] -- Efalizumab in psoriasis clinical practice: efalizumab induction and maintenance therapy for moderate to severe plaque-type psoriasis / Maria P. Stefanidou ... [et al.] -- Mitogen-activated protein kinases and their inhibitors: friend or foe in fighting psoriasis? / Stephen Hsu, Douglas Dickinson -- Etanercept (biological response modifier) modulates clinical, morphological and immunohistochemical profile of psoriatic disease, by acting on VEGF, CTACK and factor VIII cutaneous expression: open label clinical trial / Anna Campanati ... [et al.] -- Dendritic cell immunotherapy for the treatment of malignant melanoma / Filamer D. Kabigting, Doru T. Alexandrescu -- The novel cationic peptide omiganan: evaluation of antimicrobial (propionibacterium acnes) and anti-inflammatory effects / Evelina Rubinchik, H. David Friedland, Dominique Dugourd -- Malignant melanoma in ethnic skin / Filamer D. Kabigting ... [et al.] -- Malignant melanoma and its growth fraction / Gerald E. Pierard ... [et al.] -- Role of the tumor suppressor PDCD4 in the differentiation of the skin / Sachiko Matsuhashi, Takeshi Okawa, Yutaka Narisawa -- A novel transdermal drug delivery system mediated by arginine-rich intracellular delivery peptides / Betty Revon Liu ... [et al.] -- Onychomycosis caused by nondermatophytic filamentous fungi / S.F.R. Souza ... [et al.].

Subjects Acne.

	Melanoma.
	Psoriasis.
	Acne Vulgaris--therapy.
	Melanoma--therapy.
	Psoriasis--therapy.
	Therapies, Investigational.
Notes	Includes bibliographical references and index.
Series	Dermatology--laboratory and clinical research series

Human cell culture protocols
LCCN	2011939785
Type of material	Book
Main title	Human cell culture protocols / edited by Ragai R. Mitry, Robin D. Hughes.
Edition	3rd ed.
Published/Created	New York: Humana Press, c2012.
Description	xiv, 435 p.: ill. (some col.); 26 cm.
ISBN	9781617793660 (alk. paper)
	1617793663 (alk. paper)
	9781617793677 (e-ISBN)
	1617793671 (e-ISBN)
LC classification	QH585.2 .H85 2012
Related names	Mitry, Ragai R.
	Hughes, Robin D.
Contents	1. Introduction to cell culture / Christina Philippeos, Robin D. Hughes, Anil Dhawan and Ragai R. Mitry -- 2. Isolation and cultivation of dermal stem cells that differentiate into functional epidermal melanocytes / Ling Li, Mizubo Fukunaaga -- 3. Isolation, cultivation, and application of human alveolar epithelial cells / Nicole Daum, Anna Kuehn, Stephanie Hein, Ulrich F. Schaefer, Hanno Huwer, and Claus-

Michael Lehr -- 4. Culture of parathyroid cells / Peyman Björklund and Per Hellman -- 5. Functional analysis of human islets of langerhans maintained in culture / Shanta J. Persaud, Bo Liu, and Peter M. Jones -- 6. Conversion of non-endocrine human pancreatic cells to insulin-producing cells for treatment of diabetes / Min Zhao and Guo Cai Huang -- 7. Evaluation of cytochrome P450 activities in human hepatocytes in vitro / María José Gómez-Lechón, Agustín Lahoz, José V. Castell, and María Teresa Donato -- 8. Human hepatocytes: isolation, culture, and quality procedures / Daniel Knobeloch, Sabrina Ehnert, Lilianna Schyschka, Peter Büchler, Michael Schoenberg, Jörg Kleeff, Wolfgang E. Thasler, Natascha C. Nussler, Patricio Godoy, Jan Hengstler, and Andrea K. Nussler -- 9. Preclinical testing ov virotherapeutics for primary and secondary tumors of the liver / Martina Zimmermann, Timo Weiland, Michael Bitzer, and Ulrich M. Lauer -- 10. Isolation and functional studies of human fetal gastric epithelium in primary culture / Pierre Chailler, Jean-François Beaulieu, and Daniel Ménard -- 11. Isolation, characterization, and culture of normal human intestinal crypt and villus cells / Jean-François Beaulieu and Daniel Menard -- 12. Primary culture of human renal proximal tubule epithelial cells and interstitial fibroblasts / Claire C. Sharpe and Mark E.C. Dockrell -- 13. Glomerular epithelial and mesangial cell culture and characterization / Heather M. Wilson and Keith N. Stewart -- 14. Culture of isolated human adipocytes and isolated adipose tissue / Kirstin A.

Carwswell, Mi-Jeong Lee, and Susan K. Fried -- 15. Primary culture of human adipocyte precursor cells: expansion and differentiation / Thomas Skurk and Hans Hauner -- 16. Primary culture of ovarian cells for research on cell interactions in the hormonal control of steroidogenesis / Jerzy F. Galas -- 17. Human vascular smooth muscle cell culture / Diane Proudfoot and Catherine Shanahan -- 18. Culture of human endothelial cells from umbilical veins / Richard C.M. Siow -- 19. Human peripheral blood mononuclear cell culture for flow cytometric analysis of phosphorylated mitogen-activated protein kinases / Athanasios Mavropoulos, Daniel Smyk, Eirini I. Rigopoulou, and Dimitrios P. Bogdanos -- 20. Human CD4+CD25(high [superscript]) CD127 (low/neg [superscript]) regulatory T cells / Haibin Su, Maria Serena Longhi, Pengyun Wang, Diego Vergani, and Yun Ma -- 21. Human chondrocyte cultures as models of cartilage-specific gene regulation / Miguel Otero, Marta Favero, Cecilia Dragomir, Karim El Hachem, Ko Hashimoto, Darren A. Plumb, and Mary B. Goldring -- 22. Isolation and culture of human osteoblasts / Alison Gartland, Robin M.H. Rumney, Jane P. Dillon, and James A. Gallagher -- 23. Human osteoclast culture and phenotypic characterization / Ankita Agrawal, James A. Gallagher, and Alison Gartland -- 24. Effects of temperature generated from the holmium: Yag laser on human osteoblasts in monolayer tissue culture / Moustafa I. Hafez, Anne Sandison, Richard R. H. Coombs, Ian D. McCarthy, and Al-Shymaa M. Hafez -- 25. Laser microdissection microscopy: application to

	cell culture / Ahlam Mustafa, Cathy Cenayko, Ragai R. Mitry, and Alberto Quaglia -- 26. Automated adherent human cell culture (mesenchymal stem cells) / Robert Thomas and Elizabeth Ratcliffe -- 27. Culturing and differentiating human mesenchymal stem cells for biocompatible scaffolds in regenerative medicine / William R. Otto and Catherine E. Sarraf.
Subjects	Human cell culture--Laboratory manuals.
Notes	Includes bibliographical references and index.
Series	Methods in molecular biology, 1064-3745; 806 Springer protocols Methods in molecular biology (Clifton, N.J.); v. 806. Springer protocols.

MAP kinase signaling protocols

LCCN	2010935197
Type of material	Book
Main title	MAP kinase signaling protocols / edited by Rony Seger.
Edition	2nd ed.
Published/Created	New York: Humana Press, c2010.
Description	xiv, 529 p.: ill.; 24 cm.
Links	Publisher description https://www.loc.gov/catdir/enhancements/fy1612/2010935197-d.html Table of contents only https://www.loc.gov/catdir /enhancements/fy1612/2010935197-t.html
ISBN	9781607617945
LC classification	QP606.P76 M37 2010
Related names	Seger, Rony.
Subjects	Mitogen-activated protein kinases--Research--Methodology.

	Cellular signal transduction--Research--Methodology.
Notes	Includes bibliographical references and index.
Series	Methods in molecular biology; v. 661

MSKs

LCCN	2013043174
Type of material	Book
Main title	MSKs / [edited by] J. Simon C. Arthur.
Published/Produced	Austin, Texas: Landes Bioscience, [2014]
ISBN	9781587066610 (alk. paper)
LC classification	QP609.A3
Related names	Arthur, J. Simon C., 1969- editor of compilation.
Contents	Stimuli That Activate MSK in Cells and the Molecular Mechanism of Activation / Katarzyna Duda and Morten Frødin -- Role of Mitogen and Stress Activated Kinases in Histone Phosphorylation / Bojan Drobic, Beatriz Pørez-Cadahøa, Protiti Khan, Shannon Healy and James R. Davie -- The Activation of CREB Downstream of Mitogen-Activated Protein Kinase (MAPK) Signaling / Shaista Naqvi and J. Simon C. Arthur -- Role of MSKs in Regulating NF-kB Activation / Laurent L. Reber and Guy Haegeman -- MSK1 and Nuclear Receptors Signaling / Aleksandr Piskunov and Cøcile Rochette-Egly -- Regulation and Role of MSK in the Mammalian Brain / Bruno G. Frenguelli and Sonia A.L. Corrêa -- MSK1 and MSK2 Dependent Regulation of Immunity /J. Simon C. Arthur and Suzanne E. Elcombe -- The Role of MSKs in Disease / Anne Toftegaard Funding, Claus Johansen and Lars Iversen."
Subjects	Mitogen-Activated Protein Kinases.

Notes	MAP Kinase Signaling System. Includes bibliographical references and index.
Series	Molecular biology intelligence unit Molecular biology intelligence unit (Unnumbered: 2003).

Plant kinases: methods and protocols

LCCN	2011933668
Type of material	Book
Main title	Plant kinases: methods and protocols / edited by Nico Dissmeyer, Arp Schnittger.
Published/Created	New York: Humana; Springer, c2011.
Description	x, 312 p.: ill.; 27 cm.
ISBN	9781617792632 (alk. paper) 1617792632 (alk. paper) 9781617792649 (e-isbn) 1617792640 (e-isbn)
LC classification	QK898.P79 P58 2011
Related names	Dissmeyer, Nico. Schnittger, Arp.
Contents	Guide to the book plant kinases / Nico Dissmeyer and Arp Schnittger -- Age of protein kinases / Nico Dissmeyer and Arp Schnittger -- Expression and purification of active protein kinases from wheat germ extracts / Boglarka Sonkoly, Viola Bardoczy, and Tamas Meszaros -- Measurement of plant cyclin-dependent kinase activity using immunoprecipitation-coupled and affinity purification-based kinase assays and the baculovirus expression system / Hirofumi Harashima and Masami Sekine -- Mitogen-activated protein kinase activity and reporter gene assays in plants / Robert Doczi, Elizabeth Hatzimasoura, and Laszlo Bogre -- Use of

phospho-site substitutions to analyze the biological relevance of phosphorylation events in regulatory networks / Nico Dissmeyer and Arp Schnittger -- Bacterial assay to study plant sensor histidine kinases / Lukas Spichal -- Substrate analysis of Arabidopsis PP2C-type protein phosphatases / Julija Umbrasaite, Alois Schweighofer, and Irute Meskiene -- Modulating and monitoring MAPK activity during programmed cell death in pollen / Shutian Li and Vernonica E. Franklin-Tong -- Sensitizing plant protein kinases to specific inhibition by ATP-competitive molecules / Dor Salomon ... [et al.] -- Fluorescence fluctuation analysis of receptor kinase dimerization / Mark A. Hink, Sacco C. de Vries, and Antonie J.W.G. Visser -- Quantifying degradation rates of transmembrane receptor kinases / Niko Geldner -- Fluorescence correlation spectroscopy and fluorescence recovery after photobleaching to study receptor kinase mobility In Planta / Mark Kwaaitaal ... [et al.] -- Bimolecular-fluorescence complementation assay to monitor kinase-substrate interactions in vivo / Stefan Pusch, Nico Dissmeyer, and Arp Schnittger -- Chemical genetic analysis of protein kinase function in plants / Maik Bohmer, Michael Bolker, and Tina Romeis -- Modified metal-oxide affinity enrichment combined with 2D-PAGE and analysis of phosphoproteomes / Thomas Colby ... [et al.] -- Phosphoproteomics using iTRAQ / Alexandra M.E. Jones and Thomas S. Nuhse.

Subjects   Protein kinases.

Plants--enzymology.

84   Bibliography

	Phosphotransferases.
Notes	Includes bibliographical references and index.
Additional formats	Online version: Plant kinases. New York: Humana, c2011 9781617792649 (OCoLC)752194995
Series	Methods in molecular biology; v. 779 Springer protocols Methods in molecular biology (Clifton, N.J.); v. 779. Springer protocols.

Plant MAP kinases: methods and protocols

LCCN	2014939315
Type of material	Book
Main title	Plant MAP kinases: methods and protocols / edited by George Komis and Jozef Šamaj, Faculty of Science, Centre of the Region Haná for Biotechnological and Agricultural Research, Palacký University Olomouc, Olomouc, Czech Republic.
Published/Produced	New York: Humana Press, [2014]
Description	xi, 266 pages: illustrations (some color); 26 cm
Links	Table of contents http://www.loc.gov/catdir/enhancements/fy1410/2014939315-t.html Video to book http://www.springerimages.com/videos/978-1-4939-0921-6
ISBN	9781493909216 (acid-free paper) 1493909215 (acid-free paper)
LC classification	QK898.P79 P582 2014
Related names	Komis, George, editor. Šamaj, Jozef, editor.
Summary	"Mitogen-activated protein kinases (MAPKs) are versatile phosphorylating enzymes which regulate

multiple proteins involved in gene expression, cell architecture, plant development and reaction to diverse abiotic and biotic factors. The main aim of Plant MAP Kinases: Methods and Protocols is to provide established and new MAPK protocols adapted to the challenges posed by working with plants. The book contains 19 chapters which encompass a wide array of methods progressively scaling from the single gene, protein or cell level to large-scale arrays of proteomic, phosphoproteomic and interactomic data in order to uncover previously unidentified plant MAPK signaling pathways and to tackle with the challenging task of substrate identification. Techniques for MAPK sequence analysis and subcellular localization helping to identify their substrates and subcellular compartmentalization are also provided. Written in the highly successful Methods in Molecular Biology series format, chapters include introductions to their respective topics, lists of the necessary materials and reagents, step-by-step, readily reproducible laboratory protocols and key tips on troubleshooting and avoiding known pitfalls. Authoritative and practical, Plant MAP Kinases: Methods and Protocols represents a collection of useful plant MAPK protocols written by experts in the field for researchers and students." -- Back cover.

Subjects Mitogen-activated protein kinases--Laboratory manuals.
Plant proteins--Laboratory manuals.
Plants--enzymology--Laboratory Manuals.

	Gene Expression Regulation, Plant--Laboratory Manuals.
	Mitogen-Activated Protein Kinases--Laboratory Manuals.
	Plant Physiological Phenomena--Laboratory Manuals.
	Mitogen-activated protein kinases.
	Plant proteins.
Form/Genre	Laboratory Manuals.
	Handbooks, manuals, etc.
Notes	Includes bibliographical references and index.
Series	Methods in molecular biology, 1064-3745; 1171 Springer protocols
	Methods in molecular biology (Clifton, N.J.); v. 1171. 1064-3745
	Springer protocols (Series) 1949-2448

Protein tyrosine phosphatase control of metabolism

LCCN	2013945401
Type of material	Book
Main title	Protein tyrosine phosphatase control of metabolism / Kendra K. Bence, editor.
Published/Created	New York: Springer, [2013]
Description	xi, 273 p.: ill. (some col.); 24 cm.
ISBN	9781461478546 (alk. paper)
	1461478545 (alk. paper)
LC classification	QP609.P56 P774 2013
Related names	Bence, Kendra K., editor.
Summary	Tyrosine phosphorylation is a rapid and reversible protein modification catalyzed by the yin and yang activities of protein tyrosine kinases (PTKs) and protein tyrosine phosphatases (PTPs). A multitude of PTPs have been implicated in human disease, with a growing number of PTPs now

known to play major roles in prevalent metabolic diseases including obesity and type 2 diabetes. Recent studies into PTP function in the context of metabolism highlight the importance of understanding the specific substrates and binding partners of these enzymes, the regulation of PTP activity, and the cell/tissue specificity of PTP functions. This volume contains chapters which highlight many aspects of PTP function in the context of metabolism.-- Source other than Library of Congress.

Contents Redox Regulation of PTPs in Metabolism: Focus on Assays / Yang Xu, Benjamin G. Neel -- Quantitative Modeling Approaches for Understanding the Role of Phosphatases in Cell Signaling Regulation: Applications in Metabolism / Matthew J. Lazzara -- Protein-Tyrosine Phosphatase 1B Substrates and Control of Metabolism / Yannan Xi, Fawaz G. Haj -- PTP1B and TCPTP in CNS Signaling and Energy Balance / Kendra K. Bence -- PTP1B in the Periphery: Regulating Insulin Sensitivity and ER Stress / Mirela Delibegovic, Nimesh Mody -- Role of Protein Tyrosine Phosphatase 1B in Hepatocyte-Specific Insulin and Growth Factor Signaling / Águeda González-Rodríguez -- PTP1B in Obesity-Related Cardiovascular Function / Pimonrat Ketsawatsomkron -- Role of the SHP2 Protein Tyrosine Phosphatase in Cardiac Metabolism / Maria I. Kontaridis -- Metabolic Effects of Neural and Pancreatic Shp2 / Zhao He, Sharon S. Zhang -- Protein Tyrosine Phosphatase Epsilon as a Regulator of Body Weight and Glucose Metabolism / Ari Elson --

	The Role of LMPTP in the Metabolic Syndrome / Stephanie M. Stanford -- Mitogen-Activated Protein Kinase Phosphatases in Metabolism / Ahmed Lawan, Anton M. Bennett -- Glycogen Metabolism and Lafora Disease / Peter J. Roach.
Subjects	Protein-tyrosine phosphatase--Metabolism--Regulation.
	Protein Tyrosine Phosphatases--metabolism.
	Metabolic Diseases--enzymology.
	Metabolism--physiology.
	Protein-Tyrosine Kinases--metabolism.
Notes	Includes bibliographical reference and index.

Serpins and protein kinase inhibitors: novel functions, structural features and molecular mechanisms

LCCN	2009032910
Type of material	Book
Main title	Serpins and protein kinase inhibitors: novel functions, structural features and molecular mechanisms / Bojidor Georgiev and Sava Markovski, editors.
Published/Created	New York: Nova Science Publishers, c2010.
Description	xiii, 257 p.: ill. (some col.); 27 cm.
ISBN	9781607411871 (hardcover)
	1607411873 (hardcover)
LC classification	QP609.S47 S47 2010
Related names	Georgiev, Bojidor.
	Markovski, Sava.
Contents	Serpin-related diseases / Aleksandra Topic -- The roles of mammalian mitogen-activated protein kinase-activating protein kinases (MAPKAPKs) in cell cycle control / Sergiy Kostenko ... [et al.] -- Rho-kinase inhibitor in kidney disease / Toshio Nishikimi -- Targeting the epidermal growth

factor receptor pathway in glioblastoma multiforme and other intracranial malignancies / Marc-Eric Halatsch, Georg Karpel-Massler -- The serine proteinase inhibitor Z alpha-1 antitrypsin: acting on the NF-kappaB system for cytotoxicity / Matthew William Lawless -- Src family kinase inhibitors in cancer therapy / Faye M. Johnson, Gary E. Gallick -- Protein kinase inhibitors in cancer / Yiguo Hu, Shaoguang Li -- Protein kinase inhibitors in the treatment of malignant liver and kidney tumors / Panagiotis Samaras, Frank Stenner -- PAI-1 and the diet-induced obesity phenotype: background effects and inbreeding / Bart M. De Taeye, Tatiana Novitskaya, and Douglas E. Vaughan -- The structure of 1-proteinase inhibitor polymer: facts and hypothesis / Ewa Marszal -- SERPINA5 expression in the male reproductive tract is altered with advanced age / Matthew D. Anway -- Vaspin: visceral adipose tissue-derived serpin with insulin-sensitizing effects / Jun Wada -- Effect of alpha2-antiplasmin on tissue remodeling / Yosuke Kanno and Hiroyuki Matsuno.

Subjects	Serine proteinases--Inhibitors.
	Serpins--Physiological effect.
	Protein Kinases--Physiological effect.
	Serpins--physiology.
	Protein Kinase Inhibitors.
	Serine Proteinase Inhibitors--physiology.
Notes	Includes bibliographical references and index.
Series	Protein biochemistry, synthesis, structure, and cellular functions series
	Protein biochemistry, synthesis, structure, and cellular functions series.

INDEX

A

acetylation, 37, 39, 50
adipose tissue, 42, 78, 89
anti-apoptotic role, 60
antigen-presenting cell, 16
antioxidant, 59
apoptosis, viii, 2, 4, 5, 6, 9, 10, 12, 13, 14,
    16, 17, 18, 20, 21, 22, 23, 24, 25, 26, 27,
    55, 56, 59, 60, 61, 62, 63, 64, 65, 66, 67,
    69, 70, 71, 72, 73, 74
apoptotic cell death, viii, 55, 56, 58, 59, 60,
    61, 63, 64
arsenic, 58, 68
astrocytes, 18
atopic dermatitis, 75
Austria, 46
autophagy, viii, 2, 26, 57, 60, 61, 64, 66, 71,
    72

B

biological rhythms, 45, 46, 48, 51
BIX02189, 58, 59, 64
BMAL, 31, 32, 33, 34, 39

brain, 38, 39, 46

C

cadmium, v, vii, viii, 55, 56, 57, 58, 59, 60,
    61, 62, 63, 64, 65, 66, 67, 68, 69, 70, 71,
    72, 73, 74
calcium, 61, 66
cancer, 24, 44, 89
cascades, 3, 7, 15, 17, 22, 24, 26, 27, 35, 41,
    68, 69, 74
casein, 37, 51, 53
caspases, 13, 65, 69
cell biology, 44, 51
cell culture, 57, 67, 77, 80
cell cycle, viii, 2, 17, 48, 88
cell death, viii, 2, 6, 15, 17, 23, 55, 56, 57,
    58, 59, 60, 61, 62, 63, 64, 66, 69, 74, 83
cell fate, vii, ix, 23, 56, 63, 64
cell invasion, 11, 17
cell line, 6, 9, 14, 35, 57, 59, 67
cellular CLOCK, 31, 32, 35, 36, 40, 42
c-*fos*, 57, 68
Chagas disease, 11, 27

circadian clock, vii, viii, 29, 30, 31, 32, 33, 34, 35, 36, 37, 38, 40, 42, 43, 44, 45, 46, 47, 48, 49, 50, 51, 52, 53, 54
circadian clock genes, 35
circadian rhythmicity, 46, 48
circadian rhythm(s), viii, 29, 30, 31, 43, 44, 45, 46, 48, 50, 52, 54
c-JUN N-terminal kinase, viii, 30
classification, 75, 77, 80, 81, 82, 84, 86, 88
CLOCK, v, 29, 31, 32, 33, 34, 37, 39, 40, 41, 43, 46, 50, 54
cryptochrome (Cry), 31, 39, 40, 48
cues, viii, 29, 30, 38
culture, 49, 57, 77, 78, 80
cycles, 15, 38
cyclooxygenase, 16, 66
cysteine, 8, 24
cytochrome, 9, 78
cytokines, viii, 2, 3, 4, 20, 25, 58
cytotoxicity, 63, 72, 89

D

defense mechanisms, 11, 24
degradation, 39, 52, 83
degradation rate, 83
dendritic cell, 6, 7, 9, 11, 16, 18, 24, 25, 26, 27
dephosphorylation, 37
diabetes, 78, 87
diseases, 6, 15, 18, 20, 30, 87, 88
DNA, viii, 2, 45, 49, 50, 63
DNA damage, 45
DNA repair, viii, 2, 49, 50
domain structure, 26
drug delivery, 76

E

endocrine, 53, 78
endoplasmic reticulum stress (ERS), 13, 27

endothelial cells, ix, 56, 62, 72, 79
endothelium, 62, 72
environment, vii, 2, 32
environmental stimuli, 32
enzyme(s), 4, 36, 37, 84, 87
epithelial cells, viii, 14, 21, 55, 56, 60, 62, 64, 67, 68, 71, 77
epithelium, 15, 78
ER stress, 59, 60, 64, 71, 74
ERK1/2, ix, 6, 7, 8, 9, 10, 11, 12, 16, 17, 54, 55, 57, 58, 59, 60, 62, 63, 64, 69, 71
ERK5, viii, ix, 2, 54, 55, 57, 58, 59, 64, 71
evidence, viii, 5, 32, 42, 55, 57
exposure, viii, 11, 55, 56, 57, 58, 59, 60, 61, 62, 63, 64, 66, 72
extracellular signal regulated kinase(s) (ERK), vii, viii, ix, 2, 6, 11, 16, 18, 19, 20, 23, 25, 27, 30, 32, 33, 35, 40, 42, 43, 46, 49, 53, 54, 55, 56, 57, 58, 61, 63, 64, 68, 69, 70, 73, 74
extracts, 82

F

fibroblasts, 58, 78
filtration, 56, 62
flavin-containing oxidases, 36, 49

G

gene expression, 3, 5, 31, 34, 35, 48, 49, 70, 85
genes, 5, 31, 33, 34, 35, 41, 46, 47, 50, 57, 65
glomerular endothelium, 62, 72
glucose, 42, 56
glutathione, 12, 24, 26
glycogen, 37, 50
growth, viii, 2, 3, 4, 6, 14, 15, 17, 69, 76, 88
growth factor, viii, 2, 3, 14, 17, 89

Index

H

HEK293, 63, 64, 68, 74
hepatocytes, 14, 53, 63, 78
histone, 39, 54
host, vii, 2, 3, 5, 6, 7, 9, 10, 11, 12, 13, 14, 15, 19, 24
human, ix, 6, 10, 14, 17, 18, 19, 20, 24, 38, 44, 45, 51, 56, 57, 61, 62, 64, 65, 67, 68, 69, 70, 72, 73, 77, 86

I

immune response, 11, 12, 15
immune system, 6, 24
immunity, 16, 19, 23, 24
immunomodulatory, 12, 24
in vitro, 24, 38, 40, 42, 52, 65, 70, 72, 78
in vivo, 16, 24, 40, 64, 70, 72, 83
induction, 9, 13, 25, 32, 33, 41, 42, 46, 47, 50, 64, 66, 68, 70, 72, 76
infection, 6, 7, 8, 9, 10, 11, 13, 14, 16, 17, 18, 20, 22, 23
inflammation, 21, 50, 69
inhibition, viii, 2, 3, 6, 7, 8, 9, 12, 13, 14, 25, 27, 33, 43, 60, 61, 66, 69, 83
inhibitor, 7, 9, 10, 33, 43, 46, 50, 53, 54, 58, 59, 60, 61, 62, 64, 70, 71, 88
insulin, 14, 42, 53, 78, 89

J

Japan, 29, 43, 54, 55
JNK, v, vii, viii, ix, 1, 2, 3, 4, 8, 9, 10, 14, 19, 20, 21, 22, 23, 24, 25, 26, 27, 30, 35, 37, 39, 43, 52, 53, 56, 57, 58, 59, 60, 61, 62, 64, 66, 67, 69, 70, 71, 72, 73, 74

K

kidney cells, ix, 56, 58, 59, 62, 64
kidney tumors, 89
kidney(s), vii, viii, ix, 4, 55, 56, 57, 58, 59, 60, 62, 64, 65, 66, 67, 68, 88
kinase, vii, ix, 2, 3, 4, 7, 10, 11, 16, 17, 18, 19, 20, 21, 22, 23, 26, 27, 33, 37, 39, 43, 44, 45, 46, 50, 51, 52, 53, 54, 56, 57, 58, 59, 61, 63, 65, 67, 68, 69, 70, 71, 73, 74, 80, 81, 82, 88, 89
kinase activity, 43, 82

L

light, 17, 19, 30, 32, 33, 34, 35, 36, 37, 38, 40, 41, 42, 45, 46, 47, 48, 49, 50, 51, 53, 75
liver, 14, 24, 53, 54, 78, 89
LLC-PK$_1$, 57, 59, 65, 67, 71
localization, 4, 37, 39, 85
lymphocytes, 58, 63

M

machinery, 13, 30, 35, 40
macrophages, 3, 6, 7, 8, 10, 11, 12, 14, 16, 19, 25
malaria, 14, 17, 27
malignant melanoma, 76
mammalian cells, viii, 2, 49
mammals, 27, 32, 38, 39, 41, 42, 44, 46
MAPK/ERK, 18, 46, 58
MAPKs, 1, iii, v, vii, viii, ix, 2, 4, 5, 6, 8, 9, 11, 12, 13, 14, 15, 17, 29, 30, 31, 33, 35, 36, 37, 41, 43, 55, 56, 57, 58, 62, 64, 68, 84
mesangial cells, viii, 55, 57, 58, 60, 61, 64, 65, 66, 68, 72
metabolic, 87, 88

metabolism, viii, 2, 30, 43, 86, 87, 88
metals, vii, viii, 55, 58
methylation, 37
mice, 7, 8, 31, 33, 40, 45, 46, 47, 60, 61, 72
microorganisms, 6, 15
microscopy, 59, 79
mitogen, viii, 2, 4, 16, 21, 22, 25, 26, 27, 44, 52, 53, 55, 57, 67, 68, 69, 70, 73, 74, 79, 88
modifications, 37, 39, 51
molecular biology, 80, 81, 84, 86
molecules, 7, 9, 11, 12, 15, 16, 18, 36, 49, 83

N

National Academy of Sciences, 22, 24, 46, 48, 49, 70, 71
necrosis, 23, 57, 69
nervous system, 21, 32
neuronal apoptosis, 69, 74
neuronal cells, 58, 63, 73
neurons, 13, 19, 51, 74
NH2, ix, 55, 57, 67
nitric oxide, 49, 70
nitric oxide synthase, 70
North Africa, 20
NPAS2, 31, 33, 34, 41
NRK-52E, 60
nucleus, 3, 33, 38, 46

O

obesity, 25, 87, 89
organism, 30, 32
organs, vii, viii, 30, 34, 42, 49, 55
oscillation, 37, 43, 53, 54
oscillators, 30, 38
oxidation, 36, 49
oxidative stress, 19, 36, 66
oxygen, 8, 36, 41, 49, 50, 67

P

pacemaker, viii, 29, 30, 40, 51
parasite(s), vii, 2, 3, 5, 6, 7, 8, 9, 10, 11, 13, 14, 15, 16, 21, 24, 27
participants, 2, 5, 15
pathway(s), vii, viii, ix, 2, 3, 5, 6, 7, 9, 11, 12, 13, 17, 18, 19, 22, 23, 25, 26, 27, 29, 30, 31, 32, 33, 35, 36, 37, 40, 42, 43, 44, 48, 49, 50, 56, 57, 58, 59, 60, 61, 62, 63, 64, 66, 67, 68, 69, 70, 71, 72, 73, 74, 85, 89
periodicity, 30, 37, 40, 43
peripheral blood mononuclear cell, 79
phenotype, 37, 89
phosphatase, 15, 27, 51, 86, 88
phosphate, 4, 10, 16, 50, 56
phosphorylation, 3, 4, 6, 7, 8, 9, 10, 11, 14, 18, 20, 21, 25, 31, 32, 33, 36, 37, 39, 40, 51, 52, 57, 58, 59, 60, 61, 62, 63, 68, 73, 83, 86
physiology, viii, 29, 30, 31, 44, 88, 89
PI3K, 7, 10, 25, 27
PI3K/AKT, 10, 27
placenta, 16, 17
plants, 22, 49, 82, 85
proliferation, viii, 2, 6, 12, 15, 58, 59, 61, 62, 63, 72
protein kinase C, 50, 63, 73
protein kinases, ix, 2, 4, 17, 21, 23, 25, 53, 55, 57, 67, 69, 70, 76, 79, 80, 82, 84, 85, 86, 88
proteinase, 89
proteins, 4, 9, 12, 13, 24, 26, 31, 37, 44, 45, 47, 51, 52, 56, 62, 85, 86

R

radiation, 4, 17, 36, 58
reactive oxygen, 8, 36, 49, 50, 67

Index

receptor(s), viii, 2, 3, 11, 16, 17, 25, 32, 42, 46, 47, 71, 83, 89
recovery, 59, 64, 83
regenerative medicine, 80
renal podocytes, 62
renal proximal tubular cells, 57, 65, 68, 72
repair, viii, 2, 49, 50
repressor, 34, 40
residues, 3, 4, 39, 40
resistance, 15, 19
response, 4, 7, 11, 12, 13, 15, 19, 20, 22, 25, 26, 27, 36, 38, 39, 52, 60, 67, 68, 69, 72, 76
reticulum, 13, 27, 57, 66, 67
retina, 32, 33, 38, 42
rhythm, viii, 34, 43, 48, 52
rhythmicity, 33, 46, 48

S

signal transduction, 7, 23, 50, 52, 67, 81
signaling pathway, vii, viii, 2, 3, 6, 7, 11, 12, 13, 22, 27, 31, 32, 35, 36, 37, 40, 42, 44, 49, 55, 61, 64, 66, 73, 85
signalling, 21, 49, 50, 54, 71
signals, viii, 4, 29, 30, 35, 36, 38, 42, 48, 60
sleep disorders, 30, 38
species, 6, 8, 10, 36, 41, 49, 50, 67
stem cells, 27, 52, 77
stimulation, 23, 41
stress, vii, viii, 2, 4, 13, 19, 20, 22, 23, 26, 27, 36, 39, 44, 50, 52, 55, 57, 59, 60, 63, 64, 66, 69, 71, 73, 74
stress response, 19, 27, 39, 44
structure, 21, 26, 89
substrate(s), 4, 17, 20, 53, 83, 85, 87
suppression, 12, 13
suprachiasmatic nucleus, 33, 38, 46
survival, vii, viii, ix, 2, 4, 9, 10, 11, 13, 15, 29, 56, 58, 59, 60, 66, 72, 74
SV40-Mes13, 61, 64
synchronization, 30, 41, 42
synchronize, 34, 42
syndrome, 30, 38, 51
synthesis, 4, 10, 12, 25, 32, 43, 89

T

T cell, 12, 24, 79
target, 5, 18, 40, 41, 56, 57
temperature, 42, 43, 79
therapy, 72, 75, 77, 89
threonine, 2, 3, 39
tissue, 3, 4, 5, 17, 42, 53, 61, 79, 87, 89
toxicity, 56, 63, 65, 74
transcription, 3, 4, 20, 31, 32, 35, 36, 41, 44, 45, 53, 60
transcription factors, 20, 41, 45, 60
transduction, 7, 11, 21, 23, 32, 50, 52, 67, 81
treatment, 9, 15, 43, 59, 60, 61, 62, 63, 75, 78, 89
tyrosine, 3, 6, 39, 86, 87, 88

U

U0126, 58, 60, 61, 63, 64
UV irradiation, 39
UV light, 19

V

vertebrates, vii, viii, 30, 38, 49
villus, 78

Z

Zebrafish, 34, 35, 36, 41, 47
zinc, 58, 68

Related Nova Publications

FLAGELLA AND CILIA: TYPES, STRUCTURE AND FUNCTIONS

EDITOR: Rustem E. Uzbekov

SERIES: Cell Biology Research Progress

BOOK DESCRIPTION: Motility is an inherent property of living organisms, both unicellular and multicellular. One of the principal mechanisms of cell motility is the use of peculiar biological engines – flagella and cilia.

SOFTCOVER ISBN: 978-1-53614-333-1
RETAIL PRICE: $95

UBIQUITIN PROTEASOME SYSTEM: A REVIEW AND DIRECTIONS FOR RESEARCH

EDITOR: Aldrin V. Gomes, Ph.D.

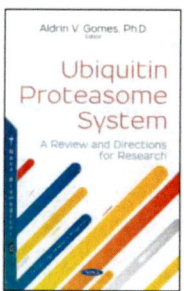

SERIES: Cell Biology Research Progress

BOOK DESCRIPTION: Over the last decade, major advancements in our understanding of the ubiquitin-proteasome system (UPS) have occurred. This book focuses on recent trends in the UPS. The UPS is possibly the most complex of all intracellular pathways – as close to 7% of all genes in the human genome make up part of the UPS.

HARDCOVER ISBN: 978-1-53613-518-3
RETAIL PRICE: $195

To see a complete list of Nova publications, please visit our website at www.novapublishers.com

Related Nova Publications

CALMODULIN: STRUCTURE, MECHANISMS AND FUNCTIONS

EDITOR: Vahid Ohme

SERIES: Cell Biology Research Progress

BOOK DESCRIPTION: In *Calmodulin: Structure, Mechanisms and Functions*, the authors consider small and poorly-studied groups of plant calcium-dependent protein kinases that directly interact with calmodulin molecules.

SOFTCOVER ISBN: 978-1-53614-948-7
RETAIL PRICE: $82

BETA-GALACTOSIDASE: PROPERTIES, STRUCTURE AND FUNCTIONS

EDITOR: Eloy Kras

SERIES: Cell Biology Research Progress

BOOK DESCRIPTION: In *Beta-Galactosidase: Properties, Structure and Functions*, the authors discuss the main microorganisms that produce β-galactosidase, the characteristics of the culture media, bioprocessing parameters, the most relevant downstream steps used in the recovery of microbial β-galactosidase, as well as the main immobilization techniques.

SOFTCOVER ISBN: 978-1-53615-605-8
RETAIL PRICE: $95

To see a complete list of Nova publications, please visit our website at www.novapublishers.com